MUSLIMS IN URBAN INDIA

About the Editors

Abdul Shaban is Associate Professor at the Centre for Development Studies, Tata Institute of Social Sciences, Mumbai. He has authored the book *Mumbai: Political Economy of Crime and Space* (Orient Blackswan, 2010), and edited *Lives of Muslims in India: Politics, Exclusion and Violence* (Routledge, 2012). He has published several papers on development and deprivation in India, and production of ethnic spaces in cities. He has authored state reports on Socio-economic and Educational Status of Muslims in Maharashtra and prepared Multi-sectoral Development Plan for Malegaon Town, for Maharashtra State Minorities Commission, Government of Maharashtra. He has also produced several other reports for Government of Maharashtra, Government of India, the World Bank, and national and international corporate groups. He has been visiting Professor at Department of Geography, University of Paris-7, Diderot, Paris, France; Department of Environmental Studies, Masaryk University, Brno, Czech Republic; International Development Studies, Palacky University, Olomouc, Czech Republic. He has been visiting fellow at London School of Economics and Political Science (LSE) in 2008-09 (at Department of Geography and Environment), and 2011-12 (at Cities Programme).

Saqib Khan is a researcher based in Mumbai. He has an MA in Development Studies from Tata Institute of Social Sciences, Mumbai. He was part of the team that prepared Multi-sectoral Development Plan for Malegaon for Maharashtra State Minorities Commission, Government of Maharashtra. He also frequently writes in online magazines about the nature and implications of present-day politics and development in India.

MUSLIMS IN URBAN INDIA

DEVELOPMENT AND EXCLUSION

Editors
Abdul Shaban
and
Saqib Khan

CONCEPT PUBLISHING COMPANY PVT. LTD.,
NEW DELHI-110059

ISBN-13: 978-81-8069-936-8

First Published 2013

Published and Printed by

Concept Since 1974

Concept Publishing Company Pvt. Ltd.
A/15-16, Commercial Block, Mohan Garden
New Delhi-110059 (India)
T: 25351460, 25351794, *F*: 091-11-25357109
E: publishing@conceptpub.com, W: www.conceptpub.com
Editorial Office: H-13, Bali Nagar, New Delhi-110 015, India.

Cataloging in Publication Data--*Courtesy:* D.K. Agencies (P) Ltd. <docinfo@dkagencies.com>

Muslims in urban India : development and exclusion / editors, Abdul Shaban and Saqib Khan.
p. cm.
Contributed articles.
Includes bibliographical references and index.
ISBN 9788180699368

1. Muslims—India—Social conditions. 2. Muslims—India—Economic conditions. 3. Marginality, Social—India. 4. Sociology, Urban—India. I. Shaban, Abdul. II. Khan, Saqib (Researcher)

DDC 305.6971054 23

Acknowledgements

We express our gratitude to our colleagues and friends without whose help and support the publication of this volume would have not been possible. Special thanks are due to Professor R.N. Sharma, Professor C. Sengupta, and Professor S. Siva Raju from Centre for Development Studies, Tata Institute of Social Sciences, Mumbai, for their assistance and encouragement at various stages of editing of this book. In many ways they made it possible that the first editor is free from other academic responsibilities so could devote his time to the volume.

We also express our sincere thanks to Dr. Ritambhara Hebbar, Dr. Anita Rath, Dr. P. Gopinath, Dr. Shailesh Kumar Darokar, Dr. Manish Kumar Jha, and Dr. P.K. Shajahan for their help and support.

Professor S. Parasuraman, Director, Tata Institute of Social Sciences has always been a source of support and encouragement. His concerns towards the marginalised communities of India are appreciable. He has made considerable efforts to mainstream the research and teaching on religious minorities and other marginalised sections within Tata Institute of Social Sciences, Mumbai, and has even gone ahead to initiate development action plan for a minority concentrated area like M-East Ward of Mumbai (Govandi-Mankhurd area), which is not only underdeveloped in various respects but is also plagued by a very high infant mortality rate.

Shaban
Saqib

Contents

Acknowledgements v
List of Tables ix
List of Figures xi
List of Box xi
Contributors xiii

Introduction 1
Abdul Shaban and *Saqib Khan*

1. Neoliberalism, Crime and Cities: Muslims, Marginalised, and Socio-spatial Praxis in Contemporary Times 18
Abdul Shaban

2. Being Muslim: Lived Space and Identity 41
Sujata Sriram and *Smriti Vaid*

3. Muslims in Urban Ghettoes in Maharashtra: Life in Deprived Neighbourhoods 62
Abdul Haleem Kidwai

4. Journey towards Educational Equality: Enrolment Patterns of Muslim Girls in Mumbai and Surrounding Towns 82
Suniti Nagpurkar

5. Poverty, Unemployment and Housing Distress among Muslims in Mumbai 123
Abdul Shaban

6. Encountering Socio-spatial Exclusion: The Experiences of Muslims in Kolkata 152
Sohel Firdos

7. Patriarchy, Purdah and Accumulation: A Study of Chikan Industry and Muslim Workers of Lucknow 175
Paaritosh Nath

8. A Peek into a Town called Malegaon 209
Saqib Khan

9. Building Inter-Community Harmony: Interventions and Practices in Mumbai 226
Ruchira Goswami

Index 244

List of Tables

1: Distribution of Workers (age-group of 15-64 years) in Different Activity Status (Principal and Subsidiary) by Socio-religious Category at All-India Level, 2004-05 7

2 : Percentage of Population (above 17 years of age) of Different SRCs Completing at Least Matric Level of Education in Select States, 2001 11

3 : Incidence of Poverty among Muslims and other Socio-Religious Categories in Urban Areas by State 12

1.1: Inmates in Jails in India 25

1.2: Socio-religious Group-wise Distribution (%) of Prisoners in Jails in India 27

1.3: Share (%) of Muslim Inmates as per different Categories in Major States in India 30

4.1: Muslim Girls' Enrolment at Institute of Distance Education at First Year (F.Y.) B.A. and (F.Y.) B.Com. 89

4.2: Enrolment of Girls in Open Category at (F.Y.) B.A. & (F.Y.) B.Com. and the share of Muslim Girls at IDE 90

4.3: Faculty-wise Enrolment of Girls at the Institute of Distance Education, Mumbai 91

4.4: Faculty-wise Distribution of the Respondents 92

4.5: Distribution of Respondents by the Education Background of the Parents 93

4.6: Faculty-wise Total Enrolment of Girls and Number of Muslim Girls in Burhani College (a Minority College) in Mumbai 101

4.7: Enrolment of Muslim Girls in the Self-Financing Courses at F.Y. Level in Burhani College (a Minority College) in Mumbai 101

4.8: Enrolment of Muslim Girls in the Self-financing Courses at 101

Non-Muslim Educational Trust College (V.E.S. College of Arts, Science, and Commerce, Chembur) at F.Y. level
4.9: Faculty-wise Enrolment of Girls in Belose College of Arts and Commerce, Dapoli 102

5.1: Per capita Monthly Family Income by Locality (per cent) 126
5.2: Per capita Monthly Family Income by Community/Sect (per cent) 128
5.3: Per capita Monthly Family Expenditure by Locality (per cent) 129
5.4: Sources of Debt by Locality (per cent) 130
5.5: Sources of Debt by Community/Sect (per cent) 131
5.6: Family Asset Ownership by Locality (per cent) 133
5.7: Occupation of Individuals by Sex 135
5.8: Share of Worker (in age-group of 15-60 years) in Various Categories of Employment/Unemployment by Sex (per cent) 136
5.9: Distribution (per cent) of Persons of age of 15 Years and above by Broad Usual Activity Status in Mumbai 137
5.10: Occupation of Individuals with Reference to Age (per cent) 139
5.11: Occupational Category by Educational Level (per cent) 143
5.12: Distribution of Sampled Families as per Locality and Type of House (per cent) 146
5.13: Distribution of Number of Persons per Room by Locality (per cent) 147

6.1: Population Characteristics of Selected Slums in Kolkata, 2001 159
6.2: Percentage of Households by Major Source of Drinking Water in Kolkata, 2001 162
6.3: Percentage of Households by Sources of Lighting, 2001 164
6.4: Percentage Distribution of Households Owning Different Assets 167
6.5: Educational Levels of the Literate Population 169
6.6: Schooling Status of Children in the Age Group of 5-19 170
6.7: Stages of Schooling at which Children Dropped Out 170
6.8: Monthly per Child Expenditure made by Parents on Education 172

7.1: Turnover of the Enterprises in the Chikan Cluster (2006-07) 179
7.2: Monthly Income of Chikankari Workers 191
7.3: Average Monthly Income of Chikankari Workers 192
7.4: Average Number of Hours Spent per day in Chikankari 193
7.5: Houses having Toilet, Water Connection and Electricity (%) 197
7.6: Share of Wages in Manufacturing Costs and Selling Price of Chikan Products 201

List of Figures

1: Share (%) of Muslim Population in the Total Urban Population by District, 2001. 3

1.1: Share of Muslims in Total Jail Inmates and Population in Major States. 29

5.1: Unemployment Rate (%) among Muslims (age-group of 15-60 years) by Locality. 137

5.2: Unemployment Rate (%) among Muslims (age-group of 15-60 years) by Sect/Communities. 138

5.3: Distribution of Families with Toilet and Bathroom within House by Locality (per cent). 148

6.1: Distribution of Slum Population in Kolkata, 2001. 160

6.2: Percentage of Slum Population in Wards of Kolkata Municipal Corporation, 2001. 161

6.3: Distribution of Households with Drinking Water Facilities. 163

6.4: Distribution of Households without Lighting Facilities, 2001. 165

6.5: Distribution of Households using Kerosene for Lighting in their Houses, 2001. 166

6.6: Age-Specific Enrolment Ratios in Kolkata. 171

List of Box

7.1: The Purdah-Poverty Nexus 184

Contributors

Abdul Haleem Kidwai is a Research Associate at the Centre for Science, Technology and Society, Tata Institute of Social Sciences, Mumbai. He has his Master degree in Development Studies from Tata Institute of Social Sciences, Mumbai. He is recipient of the Reserve Bank of India Young Scholar Award (2009). He has an interest in issues of economic development, particularly of the Muslim community. He has worked on international research projects of Brown University (US) and the Markfield Institute of Higher Education (UK).

Paaritosh Nath has a Master degree in Development Studies from Tata Institute of Social Sciences, Mumbai. He has worked extensively on Lucknow's *Chikankari* industry and was associated with the Giri Institute of Development Studies, Lucknow.

Ruchira Goswami has post-graduate degree in Social Work from Tata Institute of Social Sciences, Mumbai. She is currently working with a not-for-profit organization which is engaged in promotion of education in India.

Smriti Vaid has graduated from Tata Institute of Social Sciences, Mumbai, and has worked in capacity of Deputy Manager with New Generation Cooperative Project of National Dairy Development Board in the district of Mewat, Haryana. She has worked with milk producers of Mewat for institutional building. Currently, she works as a Project Manager in Mother NGO-Homeless project of St. Stephen's Hospital that is working with Delhi Government to improve the quality life of Homeless in

Delhi. She has also been engaged with the Homeless Survey of Delhi Government.

Sohel Firdos is Associate Professor at the Department of Geography and Natural Resources Management, Sikkim University. On completion of his Ph.D. from Centre for the Study of Regional Development, Jawaharlal Nehru University, New Delhi, Sohel joined Centre for Studies in Social Sciences, Calcutta (CSSSC) as a Fellow in Development Studies in 2002. He was a Guest Faculty for Brown International Advanced Research Institute (BIARI) in Brown University, USA, in the year 2010. He was also a Distinguished Visiting faculty to Massey University, New Zealand, in 2005 and 2007. His current works mainly relate to urban governance and socio-spatial exclusion in cities. His research articles have been published in a number of academically acclaimed journals and edited books.

Sujata Sriram is Associate Professor at the Centre for Human Ecology, Tata Institute of Social Sciences, Mumbai. Her research interests are on ethno-theories and folk beliefs, Socialisation and childcare, Work and the family, Development of the self, wellness and quality of life, Development of children in orphanages, Psycho-social counselling of children, adolescents and families, and Qualitative research—its use and applications.

Suniti Nagpurkar is Associate Professor and Head, Department of Economics at VES College of Arts, Science and Commerce, Mumbai. She has over thirty-five years of teaching experience at the undergraduate, post-graduate and management levels. She has presented research papers at state, national and international seminars. Her research interests include development economics, inclusion, gender issues and economics of education.

Introduction

Abdul Shaban and *Saqib Khan*

In the billion-plus *melting pot* of India, religious minorities are an important ingredient. Of this, Muslims with a total population of more than 138 million and a share of 13.4 per cent in the country's population are the largest religious minority group (Census of India, 2001). The history of Islam in India dates back to the seventh century AD (Elliot and Dowson, 1871). And since then it has carved out a niche in the culturally diverse population of India. However, the Muslim community has been found wanting in the economic development and employment in the country since the independence. It has also been found that the community is lagging behind in a number of other human development indicators *vis-à-vis* other Socio-Religious Categories (SRC) (Government of India, 2006; Government of India, 2007; Shaban, 2011; Shaban, 2012).

A large share of Muslim population in India lives in an adverse economic situation. In rural areas, they have higher landlessness and a majority of those owning land have lower size of landholding (NSSO, 2007). The lower educational attainments in comparison to other socio-religious groups (Government of India, 2006: 64) and prejudices against the community mar even the aspirations of those highly educated. Among others, the failing rural economy, aspiration for new lifestyle, needs for cash and better wage employments to meet the educational, health and other exigencies forces them to migrate to urban centres relatively in higher proportion to their number than any other socio-religious groups. The increased religious polarization of communities in rural areas in many states and fear of riots are also acting as push factors for

Muslims to migrate to urban areas. In urban areas, this set of national population is further confronted with more strong and intensified forms of socially exclusionary practices and violence, leading to their concentration in small areas (marginal and peripheral areas), giving rise to ghettoes (Shaban, 2008; Shaban, 2010; Robinson, 2005; Masselos, 1994; Phadke, 2007). The section of Muslims living in older parts of the towns/cities also find it difficult to move to newly developed areas due to fear of riots, lack of purchasing power and non-availability of easy housing finance. The consequence is that in urban areas a large section of Muslim population lives in slums and shanties, or dilapidated older parts of towns/cities. The poor access to institutional credit facilities and Government welfare programmes as well as disorganization within, keep the community in a vicious circle of poverty.

Muslims are distributed almost all-over the country. Apart from Jammu and Kashmir which has 67 per cent of its population as Muslims, other states like Assam (30.9 per cent of its total population), West Bengal (25.2%), Kerala (24.7%), Uttar Pradesh (18.2%) and Bihar (15.9%) have significant share of Muslim population. In fact, three states, namely Uttar Pradesh (31 million), West Bengal (20 million) and Bihar (14 million) together constitute 47 per cent of the total Muslim population in the country. A distinctive feature of Muslim community has been that the level of urbanisation of the community has been higher than the total population. In 2001, 35.7 per cent of the Muslim population was urban as compared to 27.8 per cent of the overall population. In states like Tamil Nadu, Maharashtra, Karnataka, Andhra Pradesh, Madhya Pradesh, Gujarat, and Chhattisgarh, a majority of Muslims live in urban areas. Thus, overall, Muslims are more urbanised than the general population in India (Government of India, 2006). Besides in Jammu and Kashmir, Muslims also constitute a significant share of urban population in many districts of the country, particularly in Uttar Pradesh, Bihar, West Bengal, Rajasthan, Maharashtra, Karnataka and Kerala (see Figure 1).

Muslims in India are differentiated into various groups and sub-groups along ethnic, social and cultural lines and are organised in a stratified social order (Mondal, 2003; Govt. of India, 2006; Basant and Shariff, 2010). It has been recognised that one of the foremost differentiation in the community is that between '*ashrafs*'

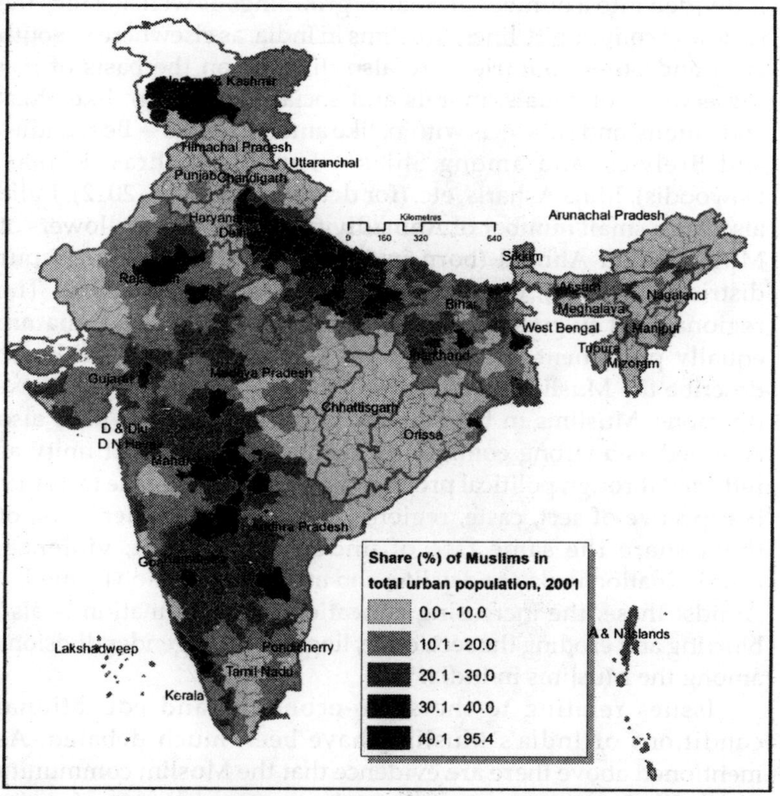

Fig. 1: Share (%) of Muslim Population in the Total Urban Population by District, 2001.

Source: Based on data from Census of India 2001.

and '*ajlafs*'. *Ashraf*, meaning noble, includes descendants of foreign Muslims (Arabs, Persians, Afghans, etc.) and converts from the higher castes of Hindus. *Ajlaf*, meaning degraded or unholy, includes the ritually clean occupational groups and low ranking converts. There is also a third category referred to as the *arzals*—converts from the erstwhile untouchable castes and who are at the bottom of the social ladder involved in occupations like scavenging, sweeping, etc. Thus, Muslims belonging to *ajlaf* and *arzal* sections have traditionally been treated as 'low castes' in their occupation and in matters of social relations (Ahmed, 1973; Siddiqui, 1994; Mondal, 1994, 2003). Each of these major divisions is divided into a number of smaller groupings as well. Besides the caste and community lines, Muslims in India, as elsewhere in South Asia and other countries, are also divided on the basis of *fiqh* (observance of rituals, morals and social legislation), like Shias and Sunnis, and sub-sects within, like among Sunnis— Beobandies and Brelvies, and among Shias— Ismailis (Bohras, Khojas, Dawoodis), Ithna Asharis, etc. (for details see Shaban, 2012). India also has a small number of Ahmadiyya Muslims, the followers of Mirza Gulam Ahmad (born in 1839/40 at Qadian, Gurdaspur district of Punjab), the founder of Ahmadiyya movement. The regional and linguistic divisions among Muslims in India are equally prominent and often used as meaningful identities to describe the Muslims in India (see Shaban, 2012). With all these divisions, Muslims in the country in the recent years have also emerged as a strong community, displaying a sense of unity as reflected through political process, mainly because of the facts that irrespective of sect, caste, region, language and gender, most of them share the same fate of underdevelopment, violence, discrimination in day to day life, and misrepresentation by media. Amidst these, the increasing education and information is also blurring and eroding the sect, caste, linguistic and gender divisions among the Muslims in India.

Issues relating to the socio-economic and educational conditions of India's Muslims have been much debated. As mentioned above there are evidence that the Muslim community lags behind other communities in the country in terms of socio-economic and educational development as well as political representation. After the mutiny of 1857, the British attempted to suppress Muslims as it was widely believed that Muslims were responsible for the mutiny and wanted to end their rule over the country. Post 1857, a considerable discrimination against and suppressions of Muslims followed. And, so the agitation of Muslims grew against the British rule, as reflected through *wahabi* movement. The attempts of the British during the years was to decentre Muslims from political power and generate the aspiration among Hindus for the same in order to break the Hindu-Muslim unity as reflected in 1857 mutiny, in which many Hindus rose against the British for a Muslim Mughal King. The British policy resulted in considerable socio-economic and political marginalisation of Muslims. A section of the British ruling class got worried with this rapid marginalisation. Then Viceroy Lord Mayo in 1870 formed a commission under Sir William Hunter to find out the reasons behind the resentment among Muslims against the British India Government and also their socio-economic situation. The Hunter Commission Report "Our Indian Musalmans" became very revealing about the discrimination and marginalisation of Muslims (see Hunter, 2002).

The partisan of the country on religious line suppressed the issues related to Muslims for quite some time after the Independence. However, constituted by then Prime Minister Indira Gandhi, the Gopal Singh Committee Report in 1983 highlighted the marginalisation and a sense of being discriminated against among Muslims (Ramakrishnan, 2006). The Committee made wide-ranging short and long-term recommendations, but it wasn't made public and could not make any headway. The Sachar Committee which submitted its report in November 2006 is the first attempt of its kind to analyse the conditions of Muslims in India using large-scale empirical data (Basant, 2007). Examining extensive data from across the country, the report held that "while there is considerable variation in the conditions of Muslims across states, the community exhibits deficits and deprivation in practically all dimensions of development" (p. 237). Basing its approach on three issues of *identity*, *security* and *equity*, the report came up with some important recommendations at both general and specific levels. Another significant step to address the concerns of the Muslim community came with the report of the National Commission for Religious and Linguistic Minorities, also known as Ranganath Misra Commission Report, in 2007. The report recommended 15 per cent reservation for minorities, including 10 per cent for Muslims, in all non-minority educational institutions and posts in all cadres and grades under the Central and State Governments (Government of India, 2007: 152).

As stated above, Muslims in India are more urbanised than the total population. Only 20 per cent of Muslims are cultivators as compared to the national average of 31.7 per cent (Government of India, 2007: 26). Muslims are, thus, less linked to land, even in rural areas. Census of India 2001 data shows that whereas among all religions about 40 per cent of rural workers were cultivators, among Muslims this figure was only 30 per cent. Agricultural workers (cultivators and agricultural labourers combined) constituted 75 per cent of total rural workers but only 60 per cent of Muslim rural workers (Government of India, 2006: 35). There is also preponderance of Muslims in self-employment in urban areas: the shares of workers in 'own account workers in household enterprise' and in 'unpaid family worker in household enterprise' are relatively much higher than the same in other SRCs (see Table 1). This is a reflection of engagement of the Muslim workers in informal sector activities and low-paid employment.

As Muslims have a substantial presence in urban areas, it is there that a large number of them come face to face with issues of *identity, security* and *equity* in myriad ways. The chapters in this book provide a glimpse into how the three issues interact and imply for Muslims in a number of urban centres in the country. Identity is one of the important concerns of the community as it (Muslim) carries a double burden of being labelled as "anti-national" and as being "appeased" at the same time (Government of India, 2006: 11). The feeling of 'suspicion' against Muslims is both at the societal level and in public sphere and spaces. This is also reflected in the way the community is marked out in public places; faces discrimination in employment and education. Over the years, the community came to be identified with acts of terror in the country in what came to be called as 'Islamic terrorism'. However, off late, the involvement of right-wing Hindutva groups in several acts of terror across the country has revealed the fallacy of identifying Muslim community *vis-à-vis* terrorism in the country. Like associating community with terrorism, the myths of appeasement, anti-nationalism, etc., many other fallacies and myths about the community are being deliberately and purposefully circulated by certain organisation not only to keep the Muslims as deprived but also to undermine the national unity and progress. Among others, one of the reasons for prevalence of myths about the Muslims in the country has been lack of objective academic studies on the community. Academic examination of religious issues and the analysis of socio-economic data on socio-religious categories have often been avoided by the mainstream researchers. Only in recent years, high impact researches with significant policy implications prominently led by Abusaleh Shariff, Rakesh Basant, Ram Puniyani, Ashghar Ali Engineer, Zoya Hasan, Mushirul Hasan, etc., have started emerging. The Sachar Committee Report, 2006, (Abusaleh Shariff and Rakesh Basant

Table 1: Distribution of Workers (age-group of 15-64 years) in Different Activity Status (Principal and Subsidiary) by Socio-religious Category at All-India Level, 2004-05

Activity Status	*Hindus*			*Muslims*	*Other Minorities*	*All*
	All Hindus	*SCs/STs*	*OBCs*			
Own account worker in household enterprise	29.7	25.5	30.9	39.4	26.9	30.9
Employer in household enterprise	2.3	0.6	1.7	2.0	4.2	2.4
Unpaid family worker in household enterprise	10.6	7.3	13.5	15.9	10.1	11.3
Regular wage/salaried	42.3	39.9	36.4	27.0	44.7	40.3
Casual workers in public work	0.1	0.2	0.1	0.2	0.1	0.1
Casual workers in other work	15.0	26.6	17.4	15.5	13.9	15.0
Total	**100.0**	**100.0**	**100.0**	**100.0**	**100.0**	**100.0**

Source: Government of India (2006: 112).

were associated with this report) also played a significant role in mainstreaming the research on the community. The young researchers, cutting across the religious divides, have also started taking interest in the issues the community is faced with. The chapters in this book are mainly contributed by young researchers and revolve around the issues of identity, security and development, the Muslim community of India is faced with.

Urban centres in India and particularly major towns and cities are undergoing significant transformation under neoliberal regime. As in rest of the world, the neoliberal policies are deepening social and economic divides in the country and exacerbat social conflicts, including religious ones. Neoliberalism contributes to the religious conflicts in the country through introduction of economic competition, withdrawal of state support, hyper-punitive practices rooted in ethnocentric policies. This creates anxieties among social groups with regard to their collective and individual interests. The first chapter for this book by Abdul Shaban attempts to explain and theorise this deepening of social divide in late capitalist phase. It also shows how the neoliberal policies intensify the social divides through selective punishment and incarceration of the weaker sections, and how planning and management of spaces, particularly in cities, have become crucial for neoliberalism, that in turn adversely impacts the social relations.

The chapter 'Being Muslim: Lived Space and Identities—Evidence from Jamia Nagar, New Delhi' by Sujata Sriram and Smriti Vaid is based on interaction with Muslim youth in Jamia Nagar area of New Delhi. Jamia Nagar is a predominantly Muslim-majority residential and business locality that grew up around the Jamia Millia Islamia University. The chapter discusses the aspects of what it means to be a Muslim, the role of religion in shaping identity, the influence of family, the impact of living in a closed environment, the role of community networks, perceptions of the majority community, and perceived discrimination emanating from interactions with others. The post-independent India has seen communal violence against Muslims in many parts of the country, mainly urban centres. It has been held that only 4 per cent of communal violence-related deaths have occurred in rural areas, while eight cities account for 45 per cent of deaths in communal violence (Varshney, 2002). Apart from physical targeting of the community, frequent riots have lead to an acute insecurity among

Muslims (Hasan, 1982). This is because of the experience of bipartisan approach of governments and law-enforcing agencies including police towards the community.

Insecurity has also lead to ghettoization of Muslim community in many cities. The search for 'safer' neighbourhoods often ends in areas that lack basic civic facilities. Govandi in Mumbai and Mumbra in Thane districts of Maharashtra are examples of such areas. Both these areas have high concentration of Muslim population and are marked by socio-economic deprivation. The chapter by Abdul Haleem Kidwai is based on a study of living conditions and socio-economic status of Muslims in Govandi (Mumbai) and Mumbra (Thane). Housing, together with lack of toilets and bathrooms, is a major concern especially in Govandi. The chapter reports cases of discrimination and lack of amenities and facilities in these Muslim concentrated areas. The larger public imagination of these areas as 'criminal areas' has made matters worse for the community. Bank credit is another front where the community faces discrimination, especially in Mumbra which has informally been marked as a 'red zone' by the banks.

While identity and security have been concerns of the Muslim community, it has also been felt that the community has not benefited from the growth process in the country (Government of India, 2006). This brings up the question of equity. The Sachar Committee Report discusses this issue with regard to Muslims for almost all dimensions of development. The report finds that Muslims rank somewhat above Scheduled Castes (SCs) and Scheduled Tribes (STs) but below Hindu-Other Backward Classes (OBCs), Other Minorities and Hindu-General (mostly upper castes) in almost all indicators considered (Government of India, 2006: 237; also see Government of India, 2007). Muslims are found to have low level of education (especially among the women the illiteracy is high) and political representation; higher rate of poverty and unemployment; lack of access to government programmes, etc. Literacy rates among Muslims are lower than most other SRCs (except SCs/STs) and are not increasing fast enough to converge with literacy rates of other groups (Basant and Shariff, 2010: 10). The total literacy rate for Muslims (59.1%, as per Census of India, 2001) is lower than the national average (64.8%), while the same among Muslim women (50.1%) is also lower than the national average for women (53.7%). The educational attainments among Muslims in urban areas are also significantly lower in comparison to other SRCs. Table 2 shows the share of population (above 17 years of age) which has completed at least matric level of education. As can be seen from the Table 2, in majority of the states share of Muslim population completing at least matric level of education is even lower than that of SCs and STs (combined). In fact, in many of the states as well as at all-India level, the share of SCs/STs completing matric level of education was lower than Muslims in 1953, but it is now higher. This shows that, over the years, the relative socio-economic status of Muslims has not improved at the rate that has been for SCs/STs (these were highly deprived communities in the country at the time of independence).

Among Muslims poverty in urban areas is much higher than other Socio-Religious Categories (SRCs). As per the estimate by Sachar Committee, in 1987-88, 1992-93 and 2004-05, the share of population below poverty line at all-India level in urban areas was 38 per cent, 33 per cent and 29 per cent, respectively, but for share of Muslim population below poverty line during the same years has been 53 per cent, 43 per cent, and 44 per cent, respectively (see Table 3). Further, the incidence of poverty among Muslims in urban areas has been higher than poverty among SCs and STs (combined) in many states, such as West Bengal, Assam, Maharashtra, Gujarat, Andhra Pradesh, Haryana, and Chhattisgarh. In fact, in states like Assam, Maharashtra, and Gujarat, the incidence of poverty among Muslims in urban areas in 1987-88 was lower than SCs/STs (combined), but in 2004-05 it has become higher. This shows that where the poverty among SCs/STs has declined at faster rate over the years, among Muslims the decline has been at very slower rate. In fact, in Maharashtra and Haryana, the share of Muslim population below poverty line has increased in 2004-05 as compared to that in 1987-88.

Lower socio-economic status of Muslims has been found to be adversely affecting women's education (Hasan and Menon, 2004). In fact, it has been held that an overarching focus on personal laws and religion has left socio-economic and gender under-examined in relation to Muslim women (Hasan and Menon, 2005). The Sachar Committee Report (2006) held that the educational status of the community is a matter of great concern, especially with regard to place of residence and gender. State level estimates

Table 2 : Percentage of Population (above 17 years of age) of Different SRCs Completing at Least Matric Level of Education in Select States, 2001

States	Year	Percentage who completed at least Matric , 2001								
		Total			Male (Urban)			Female (Urban)		
		Muslim	SCs/STs	All Others	Muslim	SCs/STs	All Others	Muslim	SCs/ STs	All Others
All-India	2001	23.9	21.1	42.5	36.1	42.1	63.0	32.2	31.8	57.9
	1953	5.4	2.1	11.0	18.3	12.8	41.0	3.2	1.8	12.4
West Bengal	2001	11.9	13.1	38.0	22.5	29.5	55.1	18.2	20.8	49.2
	1953	3.5	2.7	19.7	13.6	14.9	50.6	2.3	1.9	15.5
Kerala	2001	37.5	35.8	64.3	40.8	41.2	63.6	42.6	50.4	74.0
	1953	3.5	2.7	13.8	11.5	9.2	31.7	1.7	3.0	14.6
Uttar Pradesh	2001	17.4	16.4	37.0	28.0	35.8	65.6	26.9	25.0	61.3
	1953	4.4	2.0	9.5	14.3	11.9	43.5	2.8	1.1	13.5
Bihar	2001	16.1	11.1	31.4	43.4	41.3	68.9	32.5	19.5	54.2
	1953	5.6	2.0	9.9	25.9	13.0	43.9	3.5	1.4	8.6
Jharkhand	2001	23.1	17.4	40.7	50.8	49.3	74.3	41.4	34.4	62.8
	1953	6.3	3.2	12.2	27.7	21.3	49.8	3.3	7.2	11.1
Maharashtra	2001	35.1	32.4	52.2	36.9	50.4	63.9	36.6	39.5	60.4
	1953	7.3	2.4	11.7	18.0	15.0	43.7	3.7	1.5	15.8
Andhra Pradesh	2001	40.6	24.1	40.8	53.4	56.2	69.3	43.9	38.5	54.5
	1953	10.9	1.9	7.6	30.7	15.5	38.4	5.8	3.1	7.6
Gujarat	2001	26.1	26	41.1	33.1	40	56	25.3	32.3	54.5
	1953	6.5	2.3	10.7	17.1	13.5	37	2.3	1.7	11
Madhya Pradesh	2001	26.2	12.7	33.9	33.4	38.6	64.8	32.1	24.1	55.5
	1953	6.8	0.8	7.7	17.4	7.5	36	2.7	0.7	9.2
Tamil Nadu	2001	33.6	28.4	39.8	42.6	39.9	51	30.6	31.6	46.8
	1953	9.7	2.7	9.9	20.8	10.2	27.4	3.3	2	8.7

Source : Government of India (2006: 300).

Table 3 : Incidence of Poverty among Muslims and other Socio-Religious Categories in Urban Areas by State

States/UTs	*Total*			*Hindus*								
				All Hindus			*SCs/STs*			*Other Hindus*		
	2004-05	*1993-94*	*1987-88*	*2004-05*	*1993-94*	*1987-88*	*2004-05*	*1993-94*	*1987-88*	*2004-05*	*1993-94*	*1987-88*
1	2	3	4	5	6	7	8	9	10	11	12	13
All India	29	33	38	27	31	36	46	51	55	22	26	32
West Bengal	24	23	33	21	20	29	41	37	48	14	15	25
Kerala	23	24	45	24	25	44	41	32	61	21	24	42
Uttar Pradesh	32	35	40	27	31	33	46	57	49	22	25	30
Bihar	42	34	53	38	31	52	70	52	62	33	26	49
Assam	7	8	17	5	6	17	7	14	22	5	5	16
Jammu & Kashmir	10	5	13	9	5	14	16	7	35	7	5	10
Jharkhand	22	–	–	20	–	–	51	–	–	12	–	–
Karnataka	33	40	49	30	36	46	54	61	65	24	30	42
Uttarakhand	17	–	–	16	–	–	31	–	–	12	–	–
Delhi	21	16	16	21	15	16	46	47	44	10	5	8
Maharashtra	33	35	35	28	32	33	43	57	61	24	28	28
Andhra Pradesh	34	38	45	32	37	42	48	47	55	27	35	40
Gujarat	18	28	40	16	25	37	21	42	57	15	22	32
Rajasthan	27	31	36	26	28	35	48	47	50	18	23	29
Madhya Pradesh	48	48	43	45	47	42	69	65	67	37	40	35
Haryana	21	16	18	21	16	19	40	25	38	16	14	14
Tamil Nadu	28	40	42	29	40	43	50	57	65	25	36	40
Orissa	50	41	43	49	40	42	74	57	62	42	35	37
Himachal Pradesh	4	9	7	4	10	7	6	21	15	3	7	3
Chhattisgarh	46	–	–	47	–	–	50	–	–	45	–	–
Punjab	10	11	13	11	11	14	19	24	25	8	7	11
All other States	10	13	19	11	14	21	21	15	38	9	14	19

(*Contd.*)

(*Contd.*)

States/UTs	*Muslims*			*All Others*		
	2004-05	*1993-94*	*1987-88*	*2004-05*	*1993-94*	*1987-88*
1	14	15	16	17	18	19
All India	44	47	53	16	23	27
West Bengal	44	41	57	21	27	16
Kerala	31	27	56	12	21	39
Uttar Pradesh	43	46	58	5	7	31
Bihar	57	46	57	4	29	36
Assam	13	22	21	4	0	13
Jammu & Kashmir	11	23	13	5	0	8
Jharkhand	44	–	–	28	–	–
Karnataka	49	58	64	16	23	35
Uttarakhand	27	–	–	0	–	–
Delhi	29	30	21	2	6	5
Maharashtra	55	50	48	30	32	34
Andhra Pradesh	49	49	57	22	26	40
Gujarat	34	47	54	2	23	27
Rajasthan	39	56	47	7	14	21
Madhya Pradesh	70	60	52	7	31	39
Haryana	46	40	0	1	23	10
Tamil Nadu	24	46	44	21	34	30
Orissa	51	68	69	51	24	37
Himachal Pradesh	1	0	0	0	0	9
Chhattisgarh	62	–	–	11	–	–
Punjab	21	23	33	6	11	10
All other States	19	15	24	4	9	11

Source: Government of India (2006: 368).

suggest that the literacy gap between Muslims and the general average is greater in urban areas and for women (Government of India, 2006: 53). The community also lags behind in higher and technical education (*Ibid.*, 85). However, it has also been held that the community is increasingly looking at education as a means for improving its socio-economic status (Basant and Shariff, 2010: 10). The chapter by Suniti Nagpurkar looks into enrolment pattern of Muslim girls at undergraduate level in colleges of Mumbai and surrounding towns. The chapter highlights Muslim girls' response towards higher education. It also highlights the role of religion, parental education, and governmental policy and support etc. with regard to education of Muslim women. The chapter argues that though traditional gender roles still dominate parental expectations, there has been an increasing acceptance of women's higher education among the Muslim community.

The chapter on 'Poverty, Unemployment and Housing Distress among Muslims in Mumbai' by Abdul Shaban attempts to capture poverty and marginalisation of Muslims in employment and housing sectors in the city. It finds that a majority of Muslim families of the city live in poverty, and have poor access to credit and low asset ownership. Even in a city like Mumbai, they are engaged in subsistence occupations. Besides, Muslims are largely concentrated in slums and low class areas in the city. The chapter concludes with the argument that upliftment of Muslims demands efforts from the government on the lines that have been there for the SCs/STs in the country. Deprivation and marginalization of Muslims also has its spatial dimensions in certain cities. The chapter by Sohel Firdos uses the framework of 'socio-spatial exclusion' to understand deprivation and marginality among Muslims in Kolkata. Sohel argues that Muslims in Kolkata have been experiencing multiple disadvantages like poor living conditions, stigma, inadequate educational facilities and displacement from their traditional occupations.

The issues of equity and deprivation of Muslims have taken different forms in cities of India, especially industrial ones. At times, culture and social institutions reinforce exploitation and deprivation. The chapter by Paaritosh Nath examines issues related to Muslim *Chikankari* workers of Lucknow. *Chikan* is a white floral embroidery performed mainly on white fabric and as an art form (produced mostly by Muslim women) survives now mainly in

Lucknow. The chapter highlights the process of capitalist accumulation by exploitation of female artisans in which 'sexual division of labour' and cultural and ideological mechanisms like *purdah* play an important role. These together with the sub-contracting system ensure a precarious living for the Muslim workers. Thus, discrimination on social and religious lines coalesces with gender prejudices to produce forms of exclusion unique to women artisan groups. The chapter by Saqib Khan examines the issues Muslims are faced with in Malegaon town, a Muslim concentrated town in Maharashtra. The chapter, while highlighting the incidents of riots and bomb-blasts in the town in recent past that have strengthened communal discourses, also discusses the issues related to the lack of civic amenities, health infrastructure, problem of power-loom industry, and education and employment scenario in the town. It argues that the way politics has taken shape in the town has failed to address the concerns of the people at large and led to the weakening of the secular fabric.

In cities like Mumbai that have had a history of communal conflicts, there have also been interventions aimed at to build harmony between Hindus and Muslims. Ruchira Goswami's chapter, the last essay of this book, examines some of the harmony building interventions in the city. The chapter shows that the dominant methods of intervention include workshops, lectures and weekly group sessions with youth, and finds that religious background of interventionists does have its impact on target groups.

REFERENCES

Ahmed, Imtiaz (ed). 1973. *Caste and Social Stratification among Muslims in India*. New Delhi: Manohar.

Basant, Rakesh and Abusaleh Shariff (ed). 2010. *Handbook of Muslims in India: Empirical and Policy Perspectives*. New Delhi: Oxford University Press.

Basant, Rakesh. 2007. 'Social, Economic and Educational Conditions of Indian Muslims', *Economic and Political Weekly*, 42(10): 828-32.

Bhalotra, Sonia and Bernarda Zamora. 2010. 'Social Divisions in Education in India', in Rakesh Basant and Abusaleh Shariff (ed),

Handbook of Muslims in India: Empirical and Policy Perspectives. New Delhi: Oxford University Press.

Brass, Paul. 2003. *The Production of Hindu-Muslim Violence in Contemporary India*. Washington DC: University of Washington Press.

Census of India. 2001. *Population Census 2001*. Data on CD. Office of the Registrar General and Census Commissioner, Ministry of Home Affairs, Government of India, New Delhi.

Elliot, H.M. and John Dowson (ed). 1871. *The History of India, as Told by Its Own Historians: The Muhammadan Period*. London: Trubner Company.

Government of India (Ranganath Misra Commission Report). 2007. 'Report of the National Commission for Religious and Linguistic Minorities'. Ministry of Minority Affairs, New Delhi.

Government of India (Sachar Committee Report). 2006. 'Social, Economic and Educational Status of the Muslim Community of India', Prime Minister's High Level Committee, New Delhi.

Hasan, Zoya and Ritu Menon (ed.). 2005. *In a Minority: Essays on Muslim Women in India*. New Delhi: Oxford University Press.

Hasan, Zoya and Ritu Menon. 2004. *Unequal Citizens: A Study of Muslim Women in India*. New Delhi: Oxford University Press.

Hasan, Zoya Khaliq. 1982. 'Communalism and Communal Violence in India', *Social Scientist*, 10(2): 25-39.

Hunter, W.W. 2002. *The Indian Musalmans*. New Delhi: Rupa and Co. Ltd.

Islam, Maidul. 2007. 'The Plight of Indian Muslims: Some Missing Links in Sachar Committee Report and a Possible Alternative Model of Affirmative Action', *Social Action*, 57(3): 244-259.

Masselos, Jim. 1994. 'Postmodern Bombay: Fractured Discourses', in Sophie Watson and Kathie Gibson (eds.), *Postmodern Cities and Spaces*, pp. 199-215. Oxford: Basil Blackwell.

Mines, Mattison. 1972. 'Muslim Social Stratification in India: The Basis for Variation', *Southwestern Journal of Anthropology*, 28(4): 333-349.

Mondal, Seik Rahim. 1994. *Dynamics of Muslim Society*. New Delhi: Inter-India Publications.

Mondal, Seik Rahim. 2003. 'Social Structure, OBCs and Muslims', *Economic and Political Weekly*, 38(46): 4892-4897.

NSSO (National Sample Survey Organisation). 2007. 'Employment and Unemployment Situation among Major Religious Groups in India 2004-05, NSS 61st Round (July 2004-June 2005)', Ministry of Statistics and Programme Implementation, Government of India, New Delhi, March.

Phadke, Shilpa. 2007. 'Dangerous Liaisons: Women and Men–Risk and Reputation in Mumbai', *Economic and Political Weekly*, 42(17): 1510-18.

Ramakrishnan, Venkitesh. 2006. 'Community on the Margins', *Frontline*,

23(24). http://www.hindu.com/fline/fl2324/stories/20061215004700400.htm (accessed on 12th August, 2011).

Robinson, Rowena. 2005. *Tremors of Violence: Muslim Survivors of Ethnic Strife in Western India*. New Delhi: Sage Publications.

Robinson, Rowena. 2007. 'Indian Muslims: The Varied Dimensions of Marginality', *Economic and Political Weekly*, 42(10): 839-843.

Sengupta, Roshni. 2005. 'Communal Violence in India: Perspectives on Causative Factors', *Economic and Political Weekly*, 40(20): 2046-2050.

Shaban, A. 2011. 'Socio-Economic and Educational Status of Muslims in Maharashtra—A State Report'. Maharashtra State Minorities Commission, Government of Maharashtra, Mumbai.

Shaban, A. 2012. *Lives of Muslims in India: Politics, Exclusion and Violence*. New Delhi: Routledge.

Shaban, Abdul. 2008. 'Ghettoisation, Crime and Punishment in Mumbai', *Economic and Political Weekly*, 42(23): 68-73.

Shaban, Abdul. 2010. *Mumbai: Political Economy of Crime and Space*. Hyderabad: Orient Blackswan.

Siddiqui, M. K. A. 1974. *Muslims of Calcutta: A Study in Aspects of Their Social Organisation*. Calcutta: Anthropological Survey of India.

Varshney, Ashutosh. 2002. *Ethnic Conflict and Civic Life: Hindus and Muslims in India*. New Haven and London: Yale University Press.

Zainuddin, Sayyed. 2003. Islam, Social Stratification and Empowerment of Muslim OBCs, *Economic and Political Weekly*, Vol. xxxviii No. 46: 4898-4900.

1
Neoliberalism, Crime and Cities: Muslims, Marginalised, and Socio-spatial Praxis in Contemporary Times

Abdul Shaban

A majority of countries over the last decade or so have undergone drastic changes in relations of production. They have seen entrenchment of neoliberalism and withering away of the welfare state. The class relations have further accentuated amplifying caste, gender, religious, regional and similar divides. The divides have become visible at two levels. At the class level within communities or groups (defined on the basis of some similarities/homogeneities such as geographic, or belief) and other at national level. So, both the vertical and horizontal schisms have come to fore quite glaringly. The communities have been trying to protect their economic interest from other communities, but simultaneously within communities the upper class garners maximum benefits thus secured at the community level.

The manipulations and divisions have been led by the economic and social interests of the bourgeoisie in each of the groups, and the split and conflicts have been generated misusing and exploiting the sentiments and emotions of weaker sections by creating a false consciousness of community loyalties. Entrenchment of neoliberal policies, deepening of social and economic divides and political manipulations are also leading to rapid transformation of contemporary geographies. New geographies are being created which carry the social attributes and are graded and referred so. This, in turn, reinforces the social and spatial divisions. The emerging social conflicts manifesting in spatial contradictions, disputes, and fracas are endangering basic human rights and security of individuals. The spaces are getting more defined in 'their' and 'our' forms.

Even public spaces have become dominated by a few, and without secure public space, personal and private freedom cannot exist. One of the characteristics of a properly functioning society is that it provides a degree of security to its members. But these are the basic securities that are being jeopardised by the economic interests through social engineering and political manipulation. The world has also seen the unprecedented expansion of punishment practices that intensify social divisions rooted in class, race, religion, and caste. The gargantuanisation of cities has further expanded the divide, conflict and segregations. This has led to special emphasis on physical planning to create secure spaces and streets in cities. Theories on urban social and physical planning such as *'Defensible Space'* by Newman in 1972, and *'Broken Window'* by James Q. Wilson and George L. Kelling in 1982 are suitable examples which provide 'spatial logics' for creation of such secure spaces. The state, class and creeds have intricately been involved in creating such secure spaces from 'criminals'. In fact, criminality is a social construction, and criminals are imminently social creation and are conditioned by an array of social dynamics.

Among others, one needs, therefore, to take into account: (1) the dynamics of capitalist accumulation and that of the neoliberal project, (2) content and operation of criminal laws, (3) ethnic segregation in urban space, and (4) manufacturing of phobias to understand crime, criminality and criminogenesis in contemporary societies. The present chapter attempts to explore and explain contemporary forces behind social division and their linkages with emerging crimes and punishments in India and specifically in cities like Mumbai. The global and national issues shape the social and spatial formations at local levels. Mumbai being a major city in the country, important for capitalistic formations, is widely influenced by such forces and often in amplified ways.

Capitalist Accumulation and the Neoliberal Project

Neoliberalism exacerbates social differences. It puts economic interest above rest of the human and social issues. Neoliberal policies assert the pre-eminence of the market. They reinforce the conceptualisation of individuals as rational economic actors (Larner and Walters, 2004). State policies in the neoliberal era foreground this individuated rational actor, and seek to govern him/her as such (Herbert and Brown, 2006). The victims and oppressors both are considered rational and those who cannot comply are to be excluded, denied a space to exist in either political space (Feldman, 2004), or physical space (Mitchell, 2001). And, of course, one of the least visible spaces in which to warehouse them is the prison.

The most fundamental feature of peoples' lives is their relationship to the mode of production. Capitalism is a system of accumulation that organises production and social reproduction to extract surplus value. As it does so, accumulation generates those behaviours registered as crime by the state (Humphries and Wallace, 1980). In capitalist system, where the means of production are in private hands and where there invariably develops a division between the class that rules (the owners of the means of the production) and the class that is ruled (those who work for the ruling class), creates substantial amount of crime, often of the most violent sort, as a result of the contradictions that are inherent in the structure of social relations (Chambliss, 1975: 150). As capitalism develops, conflicts between social classes become more frequent, pronounced and violent (as a result for example, increasing proletrianisation), and more and more acts are defined as criminal (Chambliss, 1975). In the process, the cultures and social groups which stand against these formations get designated as blighted and perverted.

The strikes by the textile mill workers in Mumbai in 1980s for raising wages and against closure of the mills exemplify this. The state, colluding with owners of the mills, used its machineries to drive the workers out, which led to the penury and poverty of the families of workers resulting in entry of many youth and the mill workers in unlawful activities for survival (Shaban, 2010). The imperialist design to brand Islamic nations as antagonistic civilisation is also to be seen in the same paradigm (see Huntington, 1996; Engineer, 2011). Faced with stiff resistance from a cultural, social and religious group, the West has used all the possible tools in their hand—military, political, social and economic to demonise the nations. Further, in countries where Muslims are in minority and there is majority-minority conflict, the western design has come to help the majority in carrying out excesses on Muslims. This conflict created by neoliberal regimes, thus, is furthering many other conflicts within nations and at the international level. Some are terming this as a 'clash of civilisations'. In fact, it is nothing but a clash of economic interests (Harvey, 2003), which is not made explicit, and the war is fought in the name of civilisation, religion, ethnic groups, and terrorism.

Opponents of neoliberalism argue that neoliberalism is the implementation of global capitalism through government/military interventionism to protect the interests of multinational corporations (Pollin, 2003; Harvey, 2000; 2005). Even neoliberal proponent such as Thomas Friedman (2000) has argued approvingly that the hidden hand of the market will never work without a hidden fist. "Taken together, neoliberal policies work to entrench and deepen social inequities. One key form in which these inequities take spatial expression is in practices of exclusion, most notably in urban areas" (Herbert and Brown, 2006: 756). In Mumbai, about 78 per cent of the population live in slums and shanties (Lewis, 2011) and are often looked with division and suspicion. The people living in these slums and shanties are considered as illegal residents and are forever threatened to be thrown out or their properties destroyed/confiscated. It is a irony that these threats persist despite such people constituting majority of the city's population, indicating a democratic deficit in law making and institutions of regulations.

Spatiality of Neoliberalism

The spatiality of neoliberalism is quite discernible in the cityscape of Mumbai as also in the rest of the cities and countryside in India. In Mumbai, on the one side there are Manhattan-type skyscrapers (mainly the main city and along the coast in Western Suburb) while at the same time there are slum dwellings on thousands of acres mainly in swampy lands and garbage dumps (in Gowandi-Shivaji Nagar, Mankhurd, Dharavi). Neoliberalism also advocates 'smart growth': anti-sprawl development that is environmentally, fiscally, and economically smart and involves innovative land-use planning techniques and range of housing for providing shelter for all economic groups. Yet, since neoliberalism is based on the premise of increasing consumption and inevitably generation of wastes the logic of smart growth is put paid to. By dumping wastes in someone else's courtyard "smart growth" cannot be created. Further, given the preserve of a distorted real estate market, the question of provision of a range of housing becomes irrelevant. Seventy-eight per cent of the population living in slums in cities like Mumbai can hardly afford any respectable housing. "Housing, which otherwise should be human service, is turned into a commodity for profiteering and speculation...It is estimated that promoters and builders earn fabulous profits on housing promotion, anywhere between 100 per cent to 500 per cent on their investments. However, this booty has to be shared with politicians, bureaucrats and crime mafia as protection money. Frequent extortion killing of builders by the underworld or violent feuds among themselves are enough reflection on the sad state of affairs in the real estate business" (Sharma, 2007: 308-309). 'Smart growth' is a terminological innovation like many others created for deception. In Mumbai, resettlement and rehabilitation of the slum dwellers from railway lines and roads (that are often needed to be widened to facilitate the flow of private cars and vehicles) have been taken up by the Government of Maharashtra with the help of the World Bank loan. But the whole exercise seems to be of cornering certain groups of population at one place without even providing basic social and physical infrastructure like schools, hospitals, and transport facilities. The lack of basic amenities and job opportunities (due to odd locations) at new sites are resulting in social disorganisation and pushing youth towards crime, drug addiction and prostitution. Even those who are retaining their earlier jobs have faced erosion in their income due to increased cost of transport and payment of high electricity bills, which has been privatised recently (Shaban, 2010). Thus, poverty and vulnerability has been reintroduced into thousands of slum dwellers through the policy of providing substandard flats for the purpose of residence. Planners need to understand that spatial contexts do matter in the conduct of social action. Geography is a constituent of social behaviour and not just its backdrop (Herbert and Brown, 2006). By cornering the poor to odd locations, the state has made them more vulnerable.

The spatiality of neoliberalism gets also manifested through: (1) hyperpunitiveness, mass incarceration, and catastrophic impacts of these punitive policies on communities of disadvantage;

(2) spatialised entrenchment (neoliberalism conceals inequality while simultaneously promoting it). Thousands of citizens living in slums of Bandra and Jogeshwari in Mumbai were rounded off by the police after bomb blasts in local trains in the city on July 11, 2006 on the suspicion that some of them had participated in the act. The suspicion was based on no other ground except that these slums house largely Muslim population. Hundreds were taken to the police stations and arrested and were let off only after demonstrations and agitations by human right activists.

The social divisions are also getting mirrored in spatial ones. The expansion of and growth within cities is necessary for the health and fate of the economy dominated by capitalism. The expansion of cities is often based on the planned usurpation of lands of farmers and the capture of water resources used for subsistence by local communities. The creation and expansion of Navi-Mumbai as our extension of Mumbai was possible only due to capture of land and water resources owned by farmers and fishermen. The planned Maha Mumbai and Navi Mumbai Special Economic Zones (SEZ) on the southern edge of the Navi Mumbai to save as of 'recreation' and 'consumption' with peripheral manufacturing units are other examples of capitalist initiative to expand the city bounds and to grab land for real estate profiteering. About 1,25,000 people residing in 54 villages are going to be affected by the SEZs (Shaban and Sharma, 2005). Displaced people are bound to be affected by the vagaries of the new economy and it is but obvious that some of them would be pushed into slums and perhaps into the 'world of crime'.

Beside outer expansion, cities are also reinventing and restructuring their existing land use. Popularly, this is known as 'gentrification', through which the so-called 'blighted' areas, often the city centre and poor localities, are restructured to facilitate real estate realisation. Like any other mega cities in India, demolition of the old city for creation of newer buildings has become a reality in the city of Mumbai. The slums are being grabbed and slum dwellers displaced due to the nexus of builders, politicians and the bureaucrats. Profiteering by builders, and rent seeking by the politicians and bureaucrats has become a norm. Politicians in the city have carved of crores of rupees to now in such rent seeking (Shaban, 2010). The expansion of malls, neo-rich housing colonies and eviction of slum dwellers, are leading to shrinkage of spaces of the poor in the city. The entire operation is about not removing poverty but the poor. The socio-spatial exclusion of the poorer sections from the city is being systematically carried out. This process is stigmatised by degradation of the social, and geographic environment while the pursuit of personalised goals of profiteering remains untamed. In the pursuance of selfish goods, collective goods are being undermined and 'social traps' are being created. The consumption economy has become more central to the economic policy making. The increased gaps between the rich and the poor, the ostentatious lifestyle of the rich in contrast to the shabby existence of the poor, are resulting in increase in involvement of the youth from the poor in crimes related to property and 'quiet money' (Shaban, 2010).

Incarceration of Marginalised

The economic transformations due to neoliberalism translate into cultural anxieties that fuel exclusionary tendencies, crimes and conflicts (Herbert and Brown, 2006). To contain this, expansion of criminal justice apparatus have taken place. Imprisonment is now the principal strategy for controlling and regimenting people of certain class and origin. Police lock-ups and jails in the country are overcrowded. The occupancy rate in jails has been 128 per cent in 1999, 141 per cent in 2004 and 129.2 per cent in 2008 (National Crime Report Bureau, NCRB, 1999; 2004; 2008), and in cities like Mumbai there is hardly any room in jails to house newly arrested people (Plumber and Pathak, 2011). The national and state level data on jail capacity and inmate occupancy presented in Table 1.1 shows the sad state of affairs. The jails in all the states are overcrowded with inmates. The occupancy rates in Delhi, West Bengal, Uttar Pradesh, Gujarat, Rajasthan, Karnataka, Maharashtra, Jharkhand, and Bihar are extremely high. The annual growth rate of inmate occupancy during 1999-2008 has been as high as 8.7 per cent in West Bengal, 6.9 per cent in Karnataka, 6.3 per cent in Uttar Pradesh, 5.5 per cent in Punjab, and 5.4 per cent each in Gujarat and Rajasthan. However, during 1999-2004, the annual growth rate of jail inmates had been even higher in many states than during 1999-2009. In fact, the growth rate of jail inmates in many states is higher than the annual growth rate of their per capita real incomes. Among others, this further highlights the deepening economic

Table 1.1: Inmates in Jails in India

		1999			2004			2008				
		Available capacity	*Inmate population*	*Occupancy rate (%)*	*Available capacity*	*Inmate population*	*Occupancy rate (%)*	*Available capacity*	*Inmate population*	*Occupancy rate (%)*	*Annual growth rate (%) of inmates, 1999-2004*	*Annual growth rate (%) of inmates, 1999-2008*
1.	Andhra Pradesh	11347	12668	111.6	12270	13881	113.1	14983	14080	94.0	1.9	1.2
2.	Assam	6041	6358	105.2	6215	7828	126.0	6357	8742	137.5	4.6	4.2
3.	Bihar	26138	46168	176.6	21575	33073	153.3	31885	37709	118.3	1.9	-2.0
4.	Chhattisgarh	–	–	–	4563	8873	194.5	5219	11230	215.2	–	3.0
5.	Gujarat	5353	8284	154.8	5543	11570	208.7	6619	12279	185.5	7.9	5.4
6.	Jharkhand	–	–	–	5814	17497	300.9	12831	18403	143.4	–	0.6
7.	Karnataka	7718	8370	108.4	9271	11542	124.5	12239	13538	110.6	7.6	6.9
8.	Kerala	5659	5139	90.8	5859	7194	122.8	4409	6732	152.7	8.0	3.4
9.	Madhya Pradesh	19150	35121	183.4	17374	28932	166.5	23430	34755	148.3	1.5	-0.1
10.	Maharashtra	16621	20805	125.2	16216	24922	153.7	23345	25831	110.6	4.0	2.7
11.	Orissa	7513	10923	145.4	8006	13158	164.4	13299	14373	108.1	4.1	3.5
12.	Punjab	9569	10752	112.4	10754	15064	140.1	11897	16115	135.5	8.0	5.5
13.	Rajasthan	15653	10458	66.8	15778	13171	83.5	17627	15520	88.0	5.2	5.4
14.	Tamil Nadu	18135	16410	90.5	18152	21102	116.3	20311	16204	79.8	5.7	-0.1
15.	Uttar Pradesh	33430	51427	153.8	33868	51706	152.7	42176	80809	191.6	1.0	6.3
16.	West Bengal	19644	11360	57.8	19721	16319	82.7	20472	20230	98.8	8.7	8.7
17.	Delhi	3237	10512	324.7	5050	12610	249.7	6250	11553	184.8	4.0	1.1
	All-India	219844	281380	128.0	235012	331391	141.0	297777	384753	129.2	3.6	4.1

Note: – Shows that data are not available. The figures for Chhattisgarh and Jharkhand shown in column on annual growth rate (%) of inmates 1999-2008, are for the years 2004-08. For comparison, data for Chhattisgarh, Jharkhand and Uttarakhand have been clubbed with their parent states, Madhya Pradesh, Bihar and Uttar Pradesh, respectively. However, the data for the newly created states have also been presented.

Source: NCRB (2001, 2006, 2008).

inequity, social disorganisation, emerging criminality, and incarceration in the country.

A background check on the inmates warehoused in jails across the country is constitutive. Data presented in Table 1.2 for the year 1999, 2004, and 2008 show that people belonging to religious minorities, particularly Muslims, and those from Scheduled Castes (SCs) and Scheduled Tribes (STs) 'contribute' most of the members of inmates in Indian jails. As per Census of India 2001, Muslims constitute about 13.4 per cent of the total population in the country, but their share in total jail inmate 1999 onwards have been above 21 per cent. The share of Muslims was 24.6 per cent (28.4% in 2004) and 22.5 per cent (22.7% in 2004) of the total detenus and undertrials, respectively, in jails in 2004. The bias among the police, executive organs and politicians against Muslims has been an open secret and the global and local incidences are further reinforcing the bias and hate. Further, given the economic and social vulnerability faced by minorities, particularly Muslims, it is easy to jail them and apply tough and draconian laws like Terrorist And Disruptive Activities (Prevention) Act (TADA), 1987, the Prevention Of Terrorism Act (POTA), 2002, and Maharashtra Control Of Organised Crime Act (MCOCA) against them. In fact, these are the factors responsible for their higher share in total prisoners and the community-wise lopsided prison statistics. Almost similar are the fate of other weaker and marginalised sections — SCs and STs. Further, more than two-third of those in jails are either illiterate or educated under tenth standard in schooling. Thus, the prison statistics show enough evidence of class, caste and religion based criminality/incarceration that the neoliberal policies and cultural and religious intolerance/biases have contributed to.

In almost all the states, the share of Muslim jail inmates is higher than the share of their population in total state population. However, as shown in Figure 1.1, the difference is much higher in West Bengal, Maharashtra, Gujarat, Uttar Pradesh, Kerala, and Delhi. In West Bengal, the share of Muslims in total state population is 25.2 per cent, but the share in jail inmates was as high as 54.0 per cent in 2004, and 49.8 per cent in 2008. Maharashtra and Gujarat with 10.6 per cent and 9.1 per cent of Muslim population, had 28.0 per cent and 21.8 per cent of their respective jail inmates as Muslims in 2008. In these two states, the share of

Table 1.2: Socio-religious Group-wise Distribution (%) of Prisoners in Jails in India

Classification	*Share (%) in total population*	*As on December 31, 1999*					*As on December 31, 2004*					*As on December 31, 2008*				
		Convicts	*Undertrial*	*Detenus*	*Others*	*Total*	*Convicts*	*Undertrial*	*Detenus*	*Others*	*Total*	*Convicts*	*Undertrial*	*Detenus*	*Others*	*Total*
A. Religion-wise Distribution (%) of Jail Inmates																
Hindus	80.5	71.1	66.3	73.8	74.5	67.7	70.7	69.9	56.0	65.8	69.6	74.2	69.8	58.6	55.6	71.1
Muslims	13.4	16.5	23.2	18.1	15.9	21.4	19.1	22.5	28.4	20.3	21.5	17.6	22.7	24.6	38.0	21.1
Sikh	2.3	4.0	2.8	1.2	0.6	3.1	4.2	3.8	4.1	0.0	3.7	3.9	3.2	5.2	1.9	3.5
Christian	1.9	3.6	3.7	4.9	3.0	3.7	4.8	3.0	0.3	13.4	4.1	3.3	3.0	9.7	1.3	3.2
Others	1.9	4.8	3.9	2.0	6.0	4.1	1.2	0.8	11.2	0.4	1.0	1.0	1.3	1.9	3.3	1.2
B. Caste-wise Distribution (%) of Jail Inmates																
SC	16.2	22.0	22.5	19.9	36.5	22.4	21.2	21.6	23.4	35.4	22.0	21.0	21.8	20.3	17.0	21.5
ST	8.2	14.1	13.2	7.9	11.3	13.3	16.3	12.2	11.6	12.3	13.4	14.8	11.8	13.0	7.1	12.8
OBC	NA	32.0	34.7	32.5	21.4	33.9	29.4	26.8	28.8	30.5	27.7	30.2	30.6	38.5	13.4	30.5
Others	NA	31.8	29.6	39.8	30.8	30.4	33.1	39.4	36.2	21.8	36.8	34.0	35.8	28.2	62.5	35.2
C. Educational Standard-wise Distribution (%) of Jail Inmates																
Illiterate	45.5	37.2	35.0	29.0	54.6	35.5	38.5	34.0	38.6	31.8	36.9	30.6	35.2	29.0	36.9	33.7
Below Class X	4.1	41.7	38.9	48.6	31.3	39.8	41.0	44.2	31.7	41.8	41.9	45.3	41.9	43.4	45.2	43.0
Above Class X but below Graduate	4.1	15.9	22.1	16.9	11.8	20.4	15.7	16.1	25.7	25.3	16.2	17.4	16.3	22.5	11.5	16.7
Graduate	3.2	3.6	2.9	4.5	2.3	3.1	3.7	4.1	3.3	0.7	3.7	4.5	4.2	3.4	5.0	4.3
Post-Graduate	0.4	1.1	0.8	0.7	0.0	0.9	0.8	1.0	0.3	0.0	0.8	1.2	1.3	0.5	0.7	1.3
Diploma, Tech. Degree	0.0	0.5	0.3	0.3	0.0	0.3	0.4	0.6	0.5	0.4	0.5	1.0	1.1	1.3	0.7	1.1

Sources: 1. Share of population for various religions, castes and educational standards are taken from Census of India, 2001.

2. Inmate population has been computed for the year 1999, 2004, and 2008 using data from NCRB, New Delhi.

Muslim jail inmates was even higher in earlier years (see Figure 1.1). In Chhattisgarh, where Muslims are only about 2 per cent of the total population, the share of Muslim prisoners was as high as 7.9 per cent in 2004 and 8.2 per cent in 2008. The story is almost similar in other states. Table 1.3 further shows that in almost all the states, the share of Muslims in total convicts, undertrials, detenus and other jail inmates is substantially higher than their share in total population in the respective states. In 2008, Muslims constituted as high as 48.3 per cent (54.5% in 2004) of undertrial jail inmates in West Bengal, and 31.6 per cent (33.1% in 2004) and 19.1 per cent (29.5% in 2004) in Maharashtra and Gujarat, respectively. As shown by the Justice Rajender Sachar Committee Report (Government of India, 2006), the share of Muslims in total government (Central and state) employment is extremely low while poverty is extremely high. This shows that there has emerged terrible religious repression, bias and segregation (economic, social, political and religious), which have led to this sad state of affair.

City Planning and Spatial Fortification

Crime is a direct product of the dominant mode of production and resulting social relations. City planning, a reflection of mode of production, often shows the design prepared for marginalisation of certain section of population. To overcome crimes, architect and social planners have been recommending spatial/city designs to secure private wealth from criminals. The theories "defensible space", "broken window", "situational crime prevention", and "secured by design" are all based on the neoliberal principles and aim at privatising city space. The broken window (as implemented in New York) and other situational crime prevention theories have taken environmental aspects of criminology to an extreme. Broken window and situational crime prevention theories suggest that crimes emerge in places where informal social control is not visibly communicated, where criminogenic conditions persist within built environment (Herbert and Brown, 2006). Neighbourhoods that fail to fix broken windows display a lack of informal social control and thereby invite the criminally minded into their midst. An unrepaired broken window sends a signal that a community does not care about itself, and thus cannot defend against criminal

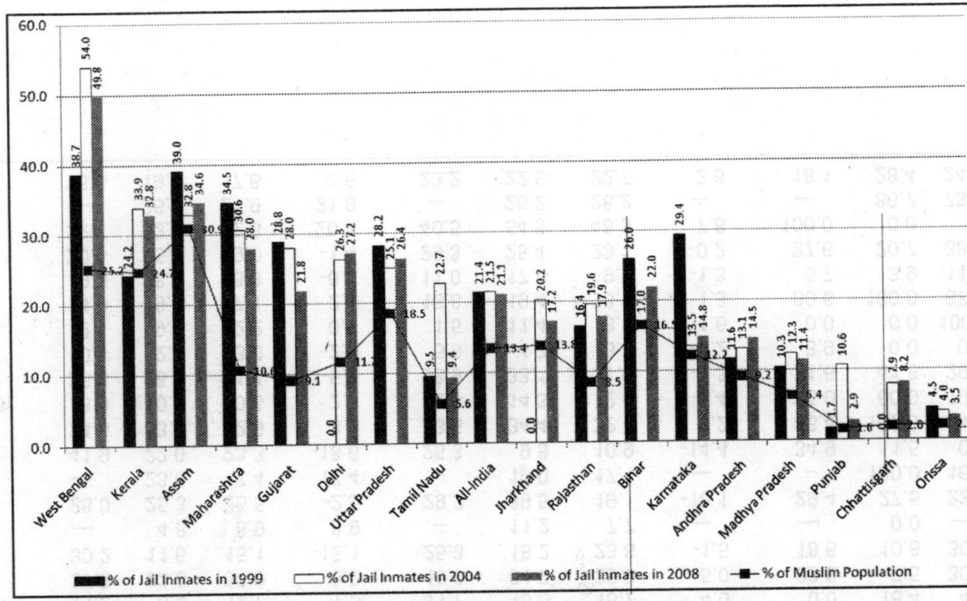

Fig. 1.1: Share of Muslims in Total Jail Inmates and Population in Major States.

Note: The figures for the newly created states, Jharkhand, Chhattisgarh and Uttarakhand have been clubbed with their parent states Bihar, Madhya Pradesh, and Uttar Pradesh, respectively, in 2004, and 2008. However, the data for newly created states are also presented.

Source: Computed using data from NCRB (2001, 2006, 2008).

Table 1.3: Share (%) of Muslim Inmates as per different Categories in Major States in India

States/UTs	*Convicts*				*Undertrials*				*Detenus*			
	1999	*2004*	*2008*	*Change 1999-2008*	*1999*	*2004*	*2008*	*Change 1999-2008*	*1999*	*2004*	*2008*	*Change 1999-2008*
Andhra Pradesh	11.2	8.9	12.7	1.5	11.7	15.3	15.7	4.0	0.0	18.4	4.8	4.8
Assam	37.2	32.3	34.0	-3.2	40.0	34.7	35.0	-5.0	23.9	7.5	30.0	6.1
Bihar	30.2	11.6	15.1	-15.1	25.3	18.2	23.8	-1.5	18.6	10.8	30.0	11.4
Chhattisgarh	—	4.8	8.9	8.9	--	11.2	7.7	—	—	0.0	—	—
Gujarat	28.0	25.3	25.5	-2.5	29.2	29.5	19.1	-10.1	29.4	27.5	23.4	-6.0
Jharkhand	—	23.0	17.4	17.4	--	19.0	17.1	—	--	100.0	16.7	--
Karnataka	41.9	22.0	23.3	-18.6	25.3	9.8	10.9	-14.4	34.9	11.5	0.0	-34.9
Kerala	24.8	33.0	32.9	8.1	23.5	34.4	32.7	9.2	86.4	100.0	41.5	-44.9
Madhya Pradesh	8.0	10.0	10.6	2.6	12.4	14.3	12.0	-0.4	25.0	60.0	19.4	-5.6
Maharashtra	26.5	26.2	21.4	-5.1	38.9	33.1	31.6	-7.3	21.6	41.8	29.2	7.6
Orissa	0.4	2.4	3.0	2.6	5.9	4.7	3.7	-2.2	3.9	0.0	0.0	-3.9
Punjab	2.1	9.0	2.2	0.1	1.5	11.4	3.1	1.6	0.0	0.0	100.0	100.0
Rajasthan	14.9	19.4	17.1	2.2	16.8	19.4	18.1	1.3	80.6	100.0	67.3	-13.3
Tamil Nadu	9.0	28.9	8.3	-0.7	11.0	17.4	9.7	-1.3	5.7	13.9	11.5	5.8
Uttar Pradesh	20.4	25.5	19.0	-1.4	29.3	25.4	29.1	-0.2	27.6	20.7	38.0	10.4
West Bengal	32.7	52.3	52.8	20.1	40.5	54.5	48.3	7.8	100.0	0.0		—
Delhi	—	26.4	21.9	21.9	--	26.2	28.2	—	—	86.7	73.7	—
All-India	16.5	19.1	17.6	2.6	23.2	22.5	22.7	2.6	18.1	28.4	24.6	2.6

(*Contd.*)

(*Contd.*)

States/UTs	*Others*				*Total*			
	1999	*2004*	*2008*	*Change 1999-2008*	*1999*	*2004*	*2008*	*Change 1999-2008*
Andhra Pradesh	19.0	0.0	8.3	-10.7	11.6	13.1	14.5	2.9
Assam	36.1	42.9		-36.1	39.0	32.8	34.6	-4.4
Bihar	26.3	8.8	19.0	-7.3	26.0	17.0	22.0	-4.0
Chhattisgarh	--	0.0	0.0	--	--	7.9	8.2	--
Gujarat		0.0	--	0.0	28.8	28.0	21.8	-7.0
Jharkhand	--	0.0	0.0	--	--	20.2	17.2	--
Karnataka	10.3	27.3	34.7	24.4	29.4	13.5	14.8	-14.6
Kerala	0.0	33.3	28.6	28.6	24.2	33.9	32.8	8.6
Madhya Pradesh	1.0	16.7	20.6	19.6	10.3	12.3	11.4	1.1
Maharashtra	15.4	13.5	--	-15.4	34.5	30.6	28.0	-6.5
Orissa	0.0	0.0	--	0.0	4.5	4.0	3.5	-1.0
Punjab	--	0.0	100.0	100.0	1.7	10.6	2.9	1.2
Rajasthan	12.9	19.0	13.8	0.9	16.4	19.6	17.9	1.5
Tamil Nadu	2.5	20.8	0.0	-2.5	9.5	22.7	9.4	-0.1
Uttar Pradesh	25.8	3.4	14.3	-11.5	28.2	25.1	26.4	-1.8
West Bengal	46.5	55.6	76.8	30.3	38.7	54.0	49.8	11.1
Delhi	--	0.0	0.0	--	--	26.3	27.2	--
All-India	15.9	20.3	38.0	2.6	21.4	21.5	21.1	2.6

Note: The figures for the newly created states, Jharkhand, Chhattisgarh and Uttarakhand have been clubbed with their parent states Bihar, Madhya Pradesh, and Uttar Pradesh, respectively, in 2004, and 2008. However, the figures for newly created states have also been presented.

Source: Computed using data from NCRB (2001, 2006, 2008).

invasion. These theories create spatial division between insiders and outsiders. The goal for each of the theories is to ensure community health through the effective repulsion of unwanted outsiders (Herbert and Brown, 2006). These theories, therefore, rely on tactics of landscape alteration for crime prevention. The theories present the semiotics of landscape in interesting fashion. They bring social construction from built environment to human behaviour and from an initial emphasis on informal social control to an endorsement of muscular policing. Valorisation of the local is a key spatial assumption of these theories. For these theories: places talk; places speak; landscapes emit messages; landscapes communicate. The broken window theory is used to justify aggressive policing and killing. The police department advocates a policy of 'zero tolerance'. Thus, hyperpunitiveness is a striking feature of neoliberalism.

The present social formation and the enduring conviction among planners and politicians about its perpetuity has led to a spatial and 'geographic reengineering' rather than performing something concrete for reducing crime. They are convinced alike that recast of spaces and architecture are the only ways to prevent crime in cities. Through the first quarter of the twentieth century, cities in the United States were developed in the form of compact, mixed-use neighbourhoods, as in European cities. That pattern began to change with the emergence of streetcar suburbs, modern architecture, zoning codes, and the ascension of the private automobile. A new system of development with a rigorous separation of uses, known as suburban development, or pejoratively as urban sprawl, arose after Second World War. The suburban working poor must spend a large portion of their incomes on transport, and the mobility of those who do not own-private car is significantly restricted in areas without good public transportation. Strip malls, auto-oriented civic and commercial buildings, and subdivisions without much individuality or character dominate the landscape. In 1970s emerged a new concept of planning called 'new urbanism'. The new urbanism is a reaction to the sprawl, based on planning and architectural principles working together to create human scale, walkable communities. It is rooted in the work of architects, planners, and theorists who believe that conventional planning thought is failing.

It is now said that housing estates built on new urbanist principles are more vulnerable to crime than private estates built in cul-de-sac format incorporating the notion of "defensible space". This is because the new urbanism opposes the cul-de-sac principles of "defensible space", and "broken window theory". It is being advocated that return to urbanism can take place only in a low-crime environment. Otherwise, people will privatise their security, retreating into cul-de-sacs and "gated communities". Crime has been seen as a major force which has forced the alteration and adoption of a new urban design. In other words, crime has always undermined urbanism. Taking 'secured by design' to its logical conclusion, everyone would live in gated, defended communities or at least ones where outsiders would not feel welcome. The city planning of Gurgaon, Noida, Navi Mumbai, high class areas of Mumbai (like Bandra, Malabar Hills), Salt Lake in Kolkata are largely based on such 'secured by design' principles contributing to emergence of "gated communities". A British police specialist Peter Knowles' report in 2003 entitled 'Designing out crime: the cost of policing new urbanism', says that new city planning fashions (new urbanism) are inherently and demonstrably criminogenic. The ideal world thus envisioned by Knowles would not have cities, towns, or village centres--places that encourage the mixing of humanity and that welcome strangers to a degree. Such a world might make Knowles' job easier, but it would be a bleak place to live. Thus, the report uses erroneous assumptions and jumps to false conclusions. The new urbanism though within the neoliberal framework is less extreme. This new urbanism is far more welcoming to 'law-abiding' strangers. The redeveloped Diggs Town in the USA feels more welcoming to outsiders because there is a clear distinction between the public and private realm, a distinction that was lacking in the old modernist design. However, the basic element these designs miss is the fact that though physical space is a factor poverty and discrimination still have the highest correlation with the 'reported' crime rates (Shaban, 2006).

Content and Operation of Criminal Law

The rules and laws are made by the bourgeois and majority groups to contain the protests and resistance by the proletariat and numerically weaker. Criminal behaviour is a product of the economic and political system (Chambliss, 1975). It first creates criminals and then tries to contain it through the application of criminal law. Criminal law is thus not a reflection of customs, but is a set of rules laid down by the 'state' in the interest of ruling class and the majority. There is a little evidence to support that the criminal law is a body of rule which reflects strongly held moral dictates of the society, but there is considerable evidence showing the critically important role played by the interest of the ruling class as a major force in the creation of criminal law. The class structure of a society acts as an important force for legal innovation. These innovations and designs are created by the ruling class in order to protect and perpetuate its position. Some hit and run cases of cars driven by the elites are suitable examples in this context. Even after killing a number of people sleeping on footpaths, the offenders wander scot-free. The same applies to the perpetrators of riots in Mumbai, Gujarat and elsewhere in the country. One with money could pay to be excused from any type or amount of crime. Similarly, an improper dance by a film actor invites appreciation, but a similar performance by a poor girl for meeting day to day needs and basics for survival is termed vulgar and banned by the state. Usurious moneylenders in villages are punished but the usurious bankers are appreciated as economic giants and pillars of social and economic development. Usurious moneylenders take the law in their hands to punish non-payers, but usurious bankers are helped by the law, and police go to arrest defaulters and confiscate their property. In India, and particularly in Mumbai, the police have been actively involved in prosecuting weaker sections and minority groups as also is the judiciary, whose decisions are based on the collected information from the police and investigating departments. Those from dominant and numerically powerful group go scot-free. In the police firing that took place at Ramabai Chawl in 1997, following the desecration of Ambedkar's statue, 11 dalits were mercilessly killed. Similarly, in riots between Hindus and Muslims in the city, the police have been taking partisan roles and had killed a significant number of Muslims. The police act in reactive instead of being a force preventive. The rioters of some political parties are allowed to create mayhem and destroy properties of millions of rupees. These facts are well brought out by Justice B. N. Shrikrishna Commission Report on the Mumbai riots in 1992-93. Several other judicial commissions on riots have also indicted leaders of majoritarian and sectarian political groups.

The July 2007 judgement of TADA court about the serial bomb blasts in Mumbai in 1993 has further created suspicion about the neutrality of the political, police and the judicial system. Those responsible for organising riots in 1992-93 in the city and killing hundreds of people from the minority groups (the death as per official figures are of 575 Muslims and 275 Hindus) are free and most of the cases against them have been dropped because of the active involvement of the state government, politicians, bureaucrats, and the police in strategy justice. Those responsible for the serial bomb-blasts and belonging to the minority community have been severely punished; a total of 12 death penalties have been awarded against members of the minority community.

Dynamics of Ethnic Segregation and Demonisation of Muslims

Several spatial pockets can be identified in cities like Mumbai, as in other cities, which are housing only members of wretched sections of society. These spaces mirror the social products and processes at work. These are the lower class, caste and minority groups of the population who generally occupy the worse off locations in the city and often these locations are marginally served by the civic amenities that modern cities otherwise are known for. Although, a mix of areas of high and low income, caste and religious groups can be found everywhere, the city of Mumbai can be divided into three main areas: (1) the main city—which largely houses the high and middle income groups and the major commercial and financial services, (2) the western suburb—the abode of business areas and residential location for high and middle income groups, (3) the eastern suburb, largely an area where the lower classes live and from where they mainly commute from to work in the main city and the western suburbs. About 70 per cent of the total population live in slums and shanties in the city. As per the City Development Plan 2005-2025 report by the Municipal Corporation of Greater Mumbai (MCGM), these slums cover only 7 per cent of the total 437.71 sq. km area of the city. Thus, there is a very high concentration and congestion in areas where the poorer section of the society live. About 72 per cent of the population in slums depend on community toilets, and about 28 per cent defecate in the open. Only 36 per cent of the population is covered by an organised system of garbage collection and clearance. About 50 per cent of the slums are covered through stand post based water supplies, 40 per cent have supply from more than one source, and remaining get from tube wells and community stand pipes. Only about 5 per cent have individual tap connections. Average monthly household income in 2001 was reported to be about Rs. 3,000 (www.mcgm.gov.in/contents/subm64ssmno128ver2-eng.pdf, accessed on 17-8-2008). Thus, a large section of the population in the city lives in sub-human condition. The lower standard of living and often high degree of economic compulsions force the lower section of population to indulge in petty crimes and sometimes in bigger crimes. The underworld dons in the city have often been from these areas and the lower section of population.

The emerging ethnic segregation of population is also quite discernible on the cityscape. In the Mumbai metropolitan region, Nagpada-Mohammad Ali Road, Jogeshwari, Kurla, Sonapur, Gowandi, Meera Road, Mumbra and Bhiwandi have emerged as significant Muslim ghettoes. These ghettos are the products of fears and biases of the majority community against the minorities. Riots have destroyed the social fabric of the city. The minorities particularly Muslims are moving to areas where they are already in comfortable number. This assures them security and the ability to practice their unhindered religious beliefs. The biases and suspicions among the majority community has been entrenched so deep by that even those sections of the Muslim population which prefer to stay in relatively developed areas are often not rented/sold flats by builders/landlords. Further, there are pockets of dalits who often are forced to live in separate areas of slums. These polarisations are reinforced by the neoliberal regime and the domination of communal politicians. The lack of confidence of one section of the society over the others creates a situation where criminality is harboured for defence of the community from other supposed rival communities. Living in a perpetual state of fear and antagonism has meant that activities such as gathering of illegal arms and ammunitions have proliferated. A charged communal atmosphere is engendered by propaganda by communal and religious organisations which also dominate the political sphere. It, therefore, is not surprising that communal clashes and tragedies such as bomb blasts keep occurring.

Manufacturing of Phobias and Fears

Communal clashes and sectarian politics have generated phobias and fears among both the majority and minority communities alike in the country. Even after the partition of the country, the Muslims are seen with same suspicion and categorised as 'other' by sectarian organisations. Secretly and openly poisons are poured into the psyche of a section of the majority community against Muslims. The propaganda against the Muslims and in recent years against the Christians as well, have started coming in open and are resulting in clashes, riots and genocides as in Gujarat in 2002. The rise of extremism in some of the Muslim countries against the occupation and exploitation of the West has further leveraged the propaganda against Muslims and their cultural practices by the various sectarian groups.

The role of media in recent years has been significant in the process of demonisation of the Muslim community and manufacturing phobias and suspicion among the majority community. In a world of concentrated wealth and major conflicts of class and *communal/religious* interest, the role of the mass media has acquired a form of systematic propaganda (Herman and Chomsky, 1988). The media, has become a tool in the hands of the powerful and dominant class to serve their ends. Profiteering and capture of viewership remain the sole aim of the media (both electronic and print), in eventually resulting in sensationalisation and dramatisation of news predominance of issues such as terrorism, riots, rape, ethnic divides, and lifestyle of glamorous artists and actors mark the content of the media. Provocative statements made by politicians and violent visuals are often repeated. The media which have become the most powerful in shaping the viewers' opinion, acts in favour of some political classes, and criticise and condemn others. The impact of the media and thrown information form bases for ideology and orientation among the population, particularly among religious majorities. If not managed and geared in proper direction, this can lead to extremely socially divisive and politically disastrous consequences. In fact, analysis shows that the amplified suspicions and hatred against the minority communities are the creation of some of the TV channels and newspapers. So, instead of promoting integration, the handling of news and their dramatisation and sensationalisation are creating social divides in the country and act towards generating and furthering the tendencies of Islamophobia and collective fear.

The elite dominate the mass media. The enormous capital required to establish an alternate source of mass media becomes a handicap for those in dissent against the dominants. Poverty, misery, lack of access to basic services to millions of population, hardly draw their attention. Provocative news, and reporting which create social divide are often repeated on screen but examples of mutual help and cooperation among communities (a tradition since the time immemorial) is hardly shown. The city spaces are also glamourised or demonised. The areas dominated by members of the high classes often depicted as safer spaces and those inhabited by the poor are presented in negative light: they are shown as areas of criminals, grime, filth, dirt, and of life-threatening. The images which are created are often far from realities. These images determine the socio-economic processes and reinforce the economic consequences. Thus, the spatial reality of cities and pervasive social constructions of reality are generated by broad array of processes. The construction of apprehension of space is not natural, inevitable occurrence, but is social creation worthy of critique (Herbert and Brown, 2006).

Conclusion

It is the relationship of individuals with the mode of production which largely determines the social outcomes like crime. In societies which are oppressive and pervaded by inequalities, poverty, and discrimination, crime is often high. The capitalistic mode of production which is based on the premise of accumulation and usurpation often generates behaviour which is termed as crime. The entrenchment of the neoliberalism in recent decades have further exacerbated the social conflicts and criminality in societies and more so in the wonders of capitalist formation – cities. Crime theories often ignore dynamics critical to urban criminality: uneven development, racial segregation, and social production of fear (Herbert and Brown, 2006: 757). In fact, without taking these into account, a complete understanding of criminality and criminal law can hardly be had.

Exclusionary tendency is core component of neoliberalism. It divides people and excludes them on the basis of economic and social achievements. The differences and schism created by it and amplified by the political class for furthering their own interest, entrenches the collective fears and suspicions of one class about other, one community about other, one caste about other, and one region about other. Neoliberalism promotes and seeks to satisfy individual objects, and defeats the formation of collective goals. It works to generate economic and cultural anxieties and fears. Though it generates growth, it also intensifies economic disparities. Mass incarceration and hyperpunitiveness are justified by neoliberal rationalities of state-society relationships and dynamics of urban capitalism (Herbert and Brown, 2006). The dynamics of (urban) capitalism, ethnic segregation, and manufacturing of phobias are complex process that transcend neighbourhood, and also critically shape the pattern of crime, much visible on cityscapes.

REFERENCES

Chambliss, William J. 1975. "Towards a Political Economy of Crime", *Theory and Society*, 2(2): 149-70.

Engineer, Asghar Ali. 2011. Prophet *of Non-Violence: Sprit of Peace, Compassion and Universality in Islam.* New Delhi: Vitasta Publsihing Pvt. Ltd.

Feldman, L. 2004. *Citizens without Shelters: Homelessness, Democracy and Political Exclusion*. New York: Cornell University Press, Ithaca.

Felson, M. 2002. *Crime and Everyday Life*, Sage Publications. California: Thousand Oaks.

Friedman, T. 2000. *The Lexus and the Olive Tree*. New York: Anchor Press.

Govt. of India (Sachar Committee Report). 2006. "Social, Economic and Educational Status of the Muslim Community of India", Prime Minister's High Level Committee, Cabinet Secretariat, New Delhi, November.

Harvey, David. 2000. *Spaces of Hope*. California: University of California Press.

Harvey, David. 2003. *New Imperialism*. Oxford University Press, New York.

Harvey, David. 2005. *A Brief History of Neoliberalism*. Oxford: Oxford University Press.

Herbert, Steve and Brown, Elizabeth. 2006. "Conception of Space and Crime in the Punitive Neoliberal City", *Antipode*, 38(7): 55-77.

Herman, Edward S. and Chomsky, Noam. 1988. *Manufacturing Consent: The Political Economy of Mass Media*. New York, Pantheon Books.

Humpheries, D. and Wallace, Don 1980. "Capital Accumulation and Urban Crime", *Social Problems*, 28(2): 179-93.

Huntington, Samuel P. 1996. *The Clash of Civilization and Remaking of the World Order*. New York: Simon & Schuster.

Larner, W. and Walters, W. 2004. "Globalisation as Governmentality", *Alternatives*, 29(5): 495-514.

Lewis, Clara. 2011. "Dharavi in Mumbai is no longer India's largest Slum", *Times of India*, Mumbai, July 6, p. 2.

MCGM. 2007. *Mumbai City Development Plan 2005-2025*. www.mcgm.gov.in/contents/subm64ssmno128ver2eng.pdf (accessed on 17th August, 2007).

Mitchell, D. 2001. "The Annihilation of Space by Law: The Roots and Implications of Antihomeless Laws in the United States", in N. Blomley and R.T. Ford (eds.), *The Legal Geographies Reader*, pp. 6-18. Malden, Massachusetts: Blackwell Publications.

NCRB (various years). "Prison Statistics—India", Ministry of Home Affairs, Government of India, New Delhi.

Pollin, Robert. 2003. *Contours of Descent: US Economic Fractures and the Landscape of Global Austerity*. New York: Verso.

Plumber, Mustafa and Pathak, Manish K. 2011. *Time to free Mumbai of its overcrowded prison?* http://www.dnaindia.com/mumbai/report_time-to-free-mumbai-of-its-overcrowded-prison_1533473 (accessed on 15th September, 2011).

Shaban, Abdul. 2006. "City, Space and Crime: A Case of Mumbai Megapolis", *Research Report*, Tata Institute of Social Sciences, Deonar, Mumbai.

Shaban, Abdul. 2010. *Mumbai: Political Economy of Crime and Space*. Hyderabad: Orient Blackswan.

Shaban, Abdul and Sharma, R. N. 2005. "Social Impact Assessment (SIA) of Maha Mumbai Special Economic Zone (MMSEZ)", *Project Report*, Tata Institute of Social Sciences, Deonar, Mumbai.

Sharma, R. N. 2007. "The Housing Market in Mumbai Metropolis and Its Relevance to the Average Citizen", in Annapurna Shaw (ed.), *Indian Cities in Transition*, pp. 383-413. Hyderabad: Orient Longman.

Wilson, J. and Kelling, G. 1997. "Broken Windows", in R. Dunham, G. and Alper (eds.), *Critical Issues in Policing*, pp. 424-437. Illinois: Waveland Press.

2

Being Muslim: Lived Space and Identity

Sujata Sriram and *Smriti Vaid*

As per the Census of India 2001, there are about 138 million Muslims in India, constituting about 13.4 per cent of the total population. Most of the Muslims in India are Sunni Muslims and there is a representation of various other Muslim sects too. The socio-economic condition of Muslims in India today, as brought out by the Sachar Committee Report (2006) reveals that socio-economically and in government services, Muslims lag behind even the Scheduled Castes (SCs) and Scheduled Tribes (STs) of the country. The history of Islam in India begins with trade and invasions by the Arabs in eighth century AD. Subsequently, large parts of north India came under Muslim rule. The growth of Islam in India was communicated through "conquest, conversion and immigration" (Herman, 2006: 68), and the Muslims in India have had a chequered past. While some Hindus did convert to Islam, the presence of Muslims in India has always been that of a minority in a land with a majority Hindu population. Communal clashes between Hindus and Muslims have taken place periodically. Issues such as cow protection have led to early incidents of Hindu-Muslim violence. The rise of Hindu fundamentalist movements, culminating in the destruction of Babri Masjid in 1992, has been flashpoints for recent violence. But in many parts of rural India, Muslims and Hindus have co-existed peacefully for generations, as demonstrated by Gottschalk (2005) in his study on Hindu-Muslim identity in a village in India. Ahmad (1993) asserts that Muslims in India have never been in social isolation, but have lived in the midst of the Hindus and other communities. In many cases, the Hindu and Muslim communities have been inextricably intertwined; while in others there has been a souring of relationships between them. The resurgence of Hindu nationalism in India has been accompanied by resurgence in Muslim nationalism (Engineer, 1991). This chapter explores the viewpoint of Muslim youth in an area of Delhi on the markers that define their identity as a minority group.

Identity

In social sciences, the terms 'identity' and 'self' are used extensively and interchangeably in many situations and with many meanings. There is a tendency for vagueness accompanied by the belief that the concept of identity will speak for itself, with a lack of emphasis on definition. Identity is of integral importance in present-day social psychology. There has been much deliberation, discussion and theorizing on issues of individual and group identity (refer to reviews by Cerulo, 1997, Frable, 1997; and Howard, 2000, for more on the subject). The popularity in the scientific and general use of the concept of identity can be attributed to the work of Erikson (2000) who used the term to represent the essence of the self as defined by membership to a particular group, as also a personal uniqueness of the self. Identity is more than just the individual; it is the relationship of the individual with the society. According to Verkuyten, identity "is about the intricacies, paradoxes, dilemmas, contradictions, imperatives, superficialities and profundities of the way individuals relate to and are related to the world in which they live" (2005: 42).

The literature on identity refers to aspects of personal and social identity, which can be seen as lying on the two ends of a continuum, with personal identity referring to the idiosyncratic features that characterise the individual, separating one from the other; and social identity largely defined by group membership. The salience of the personal and social depends largely on the context, but are difficult to separate, as aspects of one's group identity have important implications for the personal identity e.g., on issues of self-esteem. While early literature on identity showed that it was a fixed and rather immutable entity, later research and empirical evidence has shown that identity is an evolving process; and that individual identity can grow over time as a result of personal experiences and changing social contexts (Howard, 2000; Cerulo, 1997; Frable, 1997).

The first social group that children come into contact with—from which aspects of values, beliefs, roles, traditions, prejudice and stereotypes and a host of other ways of understanding the world is acquired—is the family. The family assumes great importance in the individual's socialization, for learning the importance of carriers of culture as shown by Moghaddam (2006), Verkuyten (2005) and Durkin (2004). Families themselves are nested in the wider social and cultural contexts. The family serves as the transmission mechanism of the values, beliefs, practices and rituals of the culture to children through the processes of acculturation and enculturation. It is the family that provides role models for children for learning how certain practices are transacted. It is within the context of the family that the various social bases of identity, including that of religion, assume importance and clarification.

Religion and Identity

While there have been extensive reviews in the field of social psychology on the various social bases of identity (Frable, 1997; Cerulo, 1997 and Howard, 2000), there is seldom a mention of the role of religion in the development of one's social and personal identity. Howard mentions ethnicity and race, gender and sexuality, social class, disability and age as being the social bases of identity. There is no reference to the role of religion in the development of identity. Durkin (2004) asserts that while all individuals will develop a sense of gender and ethnic identity, it may not be necessary to similarly develop a religious identity. Verkuyten (2005) says that social identities related to collective aspects such as religion, race, nationality and ethnicity are more enduring. Membership to a religious group provides the necessary cultural, ontological and structural components necessary for the development of a social identity. An examination of social psychology literature on identity shows that there has been a conflation of religious identity with ethnic and race identity, such that religious identity is rarely considered as an entity with its own merits.

Studies from the sociology of religion have tried to study the role of religion for maintaining group identity and solidarity among immigrant groups. Many of these studies draw a relationship between religion and ethnic identity (Peek, 2005). The relationship between religion and one's personal and social identities has been discussed at length by Seul (1999) and Peek (2005). Both these authors assert that religion is at the centre of an individual's personal and social identity. Religion provides vital identity supporting features such as rituals, beliefs, traditions, institutions, cosmologies and moral frameworks. It is these factors that help provide the necessary psychological bases of self-esteem, self-actualization and stability. Belonging to a religious group involves commitment of the individual, such that once such a religious identity has been formed, it becomes difficult to set aside. The impact of religion becomes consolidated through the various beliefs of the salvation of the individual and the group. Aspects of the believers as contrasted with non-believers are emphasized by religious leaders. The accompanying rituals and ceremonies increase the feeling of belonging to the group, cementing the individual to the group. Apart from defining self-identification, religion can serve as a means of group differentiation, which goes beyond other group markers.

Matsumoto and Juang (2003) emphasise the role played by religious institutions in enculturation; however most twentieth century psychologists have failed to study how religious organizations and affiliation helps in defining personal and social identity. *"In a world where religion is increasingly becoming a visible target of cross-cultural conflicts and misunderstandings, it is of utmost importance for us to continue exploring how religion defines and shapes an individual's personal experiences, belief systems and identity"* (Matsumoto and Juang, 2003:153).

Ismail (2004) reflects on the intersection of Islam and identity in the context of public and private spaces, as defined by the use of identity markers specific to the religion, which help demarcate Muslims from other religions. Significant among these markers are the *hijab* or the veil worn by women. When multiple identities are present, it becomes necessary to use what Stryker (1980) defines as identity salience in an identity hierarchy, wherein identities higher in the hierarchy assume greater importance as compared to those lower down.

Plural Societies and their Impact

When plural societies such as India are considered, various cultural groups interact with each other. When cultural groups interact with each other, there are various cultural and psychological changes that can take place, which have been referred to as acculturation. There are different strategies that various cultural groups can adopt, to deal with the process of contact. When minority and majority groups come into contact with each other, depending on the varying responses of both groups, it is possible to identify four strategies, arranged on continuum of maintenance of culture and identity and the degree of contact sought. When a minority community does not value its own identity and seeks active contact with the host or majority community, the conditions favour assimilation, as seen in the concept of the melting pot, wherein differences between cultures are ignored in favour of a large conglomerate. If on the other hand, there is a value given to preserving one's own culture and identity, accompanied by a desire for positive contact with the host community, which in turn is favourable to the minority group retaining its identity, the situation gives rise to multiculturalism and integration of both communities. Conversely, if there is a desire to maintain minority identity and there is no desire for contact with the host community, the situation that arises will be one of segregation and separation. Yet again, a disinterest in one's own identity accompanied by a disinclination to have a contact with the host community gives rise to exclusion and marginalization (Berry, 2006).

Furthermore, Berry (2006) refers to two models that can be considered in the context of plural societies. One model is that of the melting pot, where minority groups exist at the periphery of a dominant society. The assumption is that the minority groups will be assimilated within the mainstream, failure to do so results in a further marginalization of the group under question. This melting pot phenomenon assumes that the identity of minority groups is less important and subordinate to that of the mainstream society. The other model is the multicultural model, where ethno-cultural groups are allowed to and encouraged to retain their own identities and participate in society on that basis. There are shared norms for living, and societal institutions evolve to serve the interests of all the groups concerned.

The Contact Hypothesis

The nature of contact between groups can have a considerable role to play with regard to the nature and degree of segregation between communities. The contact hypothesis has been put forward to examine intergroup relations (Brewer, 2007; Hewstone, Cairns, Voci, Paolini, McLernon, Crisp, Niens and Craig, 2005; Dovidio, Gaertner, Hodson, Houlette and Johnson, 2005). The contact hypothesis suggests that if opposing groups are brought together under optimal conditions, then prejudice can be reduced and intergroup relations improved. Four conditions are necessary for reducing conflict. The first indicates that there should be equal status between the groups concerned, or at least between the individuals who interact. The second condition is that there should be cooperation between the groups in question, such that there is what Brewer (2007) refers to as cooperative interdependence. Close contact between individuals and groups such that there is a potential for acquaintanceship and intimate contact between the groups. Lastly, institutional legitimacy wherein there is the support of the authority for the development of social norms greatly facilitates contact between groups and can be used to reduce prejudice and improve intergroup relations. Hewstone *et al.* (2005) have studied how contact, both direct and extended, has helped reduce intergroup conflict in Northern Ireland. Contact in the spheres of education and community have succeeded in substantially reducing conflict between factions.

The Present Study

This chapter tries to examine the viewpoint of Muslim youth living in the city of Delhi on how they negotiate their lives as a minority group. The role of religion in defining the identity of these young people was an integral part of the study. The part played by identity markers as defined by the sample in determining their identity, as also the perceived discrimination from the majority group, along with the approach and avoidance mechanisms used in relation to the majority group are investigated. The influence of the physical space in which the youth live has a significant impact, which cannot be downplayed.

The Setting

The present study is an exploration on the identity of Muslim youth living in Delhi. The sample was drawn from families of upper middle class and middle class backgrounds of a residential colony, Jamia Nagar in Delhi. The youth drawn to take part in the study lived here with their families.

Jamia Nagar was identified as the universe for the study as it is one of the areas in Delhi where there is a higher concentration of Muslim majority population. Living in Jamia Nagar helped delineate the influence of living in a closely-knit community surrounded by others who followed the same faith and identity. Jamia Nagar as a residential area has grown around the Jamia Millia Islamia University and has long been a preferred residential and business locality for Muslims in Delhi. Jamia Nagar itself is made up of clusters of smaller localities with disparities with regard to access to resources and infrastructure. The neighbourhood caters to people of varied social classes from the lower income groups to the wealthy. It is not just a residential area; but interspersed with business clusters.

The development of Jamia Nagar has been from a small village called Okhla in the outskirts of Delhi (Soofi, 2007). In 1935, with the setting up of the Jamia Millia Islamia University, people from old Delhi, along with migrants from other parts of India moved to settle, in what is today known as Jamia Nagar. The main attraction to Jamia Nagar was the access to better education and job opportunities. Apart from the University, it is likely that Jamia Nagar represented a perseverance of Islamic culture and religion for the Muslim migrants who settled in the area. A point to note is that Jamia Nagar has been relatively free of communal violence, even when communal violence affected other parts of the Delhi city. The close association with religion in Jamia Nagar is clearly seen, with the presence of many mosques in the area. The locality today can be described as multi-ethnic, with a majority Muslim population and a small representation of other communities (*http://thedelhiwalla.blogspot.com/2007/07/jamia-nagar-delhis-rich-muslim-ghetto.html*, accessed on 3rd October, 2009).

Jamia Nagar has been referred to variously as a ghetto. However, this point is debatable. The facilities and opportunities available in the locality are such that the area is self-sufficient and little need for residents to venture out. Facilities like schools, colleges, hospitals and a market place are all available. However, there are some inherent problems that are associated with the area. Mobile phone connections are difficult to get if an address in Jamia Nagar is given. Certain services that residents from other parts of the city take for granted are unavailable here—such as pizza delivery and an ATM outlet for cash withdrawal. The area is identified as a red alert area by the police. Over the years, there have been incidents of police attacks on suspected Muslim terrorists, who have taken shelter in Jamia Nagar (Soofi, 2008; *http://thedelhiwalla.blogspot.com/2008/04/dateline-jamia-nagar-no-pizzas-for.html*, accessed on 3rd October, 2009).

In the context of lived spaces, physical segregation of residential spaces serves the functions of provision of a base for self-defence, avoidance of embarrassing contacts with unfriendly outsiders, preservation of a way of life and a base from which to attack enemies. There is, in short, safety in segregation and residential segregation increases as a direct result of large population movements in response to intimidation, as families move from religiously mixed areas into safe havens dominated by people sharing the same religious beliefs.

The Sample and Methods

The qualitative study comprised 29 young people in the age group of 18-25 years. All participants of the study were Muslims living with their families in Jamia Nagar. Sample selection was done purposively with the help of key respondents who facilitated the entry into the field and carried out the necessary introduction with people who would become participants in the study. The respondents were 18 men and 11 women from upper middle class and middle class backgrounds. If we consider their regional backgrounds, 18 respondents hailed from various parts of Uttar Pradesh, five respondents said their families had moved to Jamia Nagar from other parts of Delhi, and the rest were from various other states in India. Most respondents were second-generation residents of Jamia Nagar and it was their parents who had moved here to benefit from the reward of living and working in a large city and the additional advantage of living in a Muslim locality with security. Only one respondent said his family had been living in the locality for the last 100 years.

Of the 29 respondents who took part in the study, two worked full-time; two others held part-time jobs and were studying alongside. One respondent was looking for a job on completing her education. The other 24 respondents were all students, studying various courses in Jamia Millia University. All four working respondents (full-time and part-time) were men. It is interesting to note that most respondents had completed their schooling from various schools in Jamia Nagar itself.

The fathers of the respondents were the main earning members—some were occupied in professional pursuits as teachers and engineer; but the major source of revenue for the families seemed to be from business of some kind, usually a family business. There were two families where agricultural revenue from family lands was the main source of income. The mothers of the respondents were mainly homemakers; only three mothers worked as teachers.

In-depth interview was the method used to collect data from the respondents. An interview guide was prepared to allow investigation into aspects of perceptions of being a Muslim as a minority, the influence of the family and community about the religious practices followed, awareness about and following of identity markers of the religion, perceptions of discrimination from the majority community, political awareness and awareness of contemporary issues concerning the religious group, among other such issues. Each interview was conducted over two sessions of about two hours. Many respondents were curious about the purpose of the study especially since one of the authors, who was also collecting data, was a Hindu.

According to Johnson (2001:104), "A researcher who uses in-depth interviewing commonly seeks "deep" information and knowledge—usually deeper information and knowledge than is sought in surveys, informal interviewing, or focus groups, for example." This information usually concerns very personal matters, such as an individual's self, lived experiences, values and decisions, occupational ideology, cultural knowledge or perspective. As members of a particular culture, there is an understanding about the nuances, the cultural activities, the routines, the rituals and beliefs that accompany such membership. This understanding has to be derived by the researcher in order to appreciate what membership to a group entails. In-depth interviewing allows the scope to probe, to delve into the nuanced meaning to obtain from the respondent the various explanations that are developed to understand and negotiate the cultural space in which they dwell. In-depth interviewing allows for the "voice" of respondents to resonate, allowing them to express what they feel on issues that are of concern for them (Bhatia, 2007). A pilot study was carried out with five subjects to establish the sequence of questions in the interview guide and determine the reliability and validity of the same. Minor modifications in the wording and sequence of questions were done later.

Findings and Discussion

The Role of Religion

All 29 respondents in the study were born into families that practised the beliefs of Islam. For them, being Muslim was a given, a fact of life that was not questioned at all. Being Muslim appeared to be an assigned identity. For the respondents, being Muslim was an integral part of their lives. Their values and belief systems were defined by the religion to which they belonged, from birth onwards. Being Muslim helped in their being connected with others who were like them, emphasising the part played by religion in determining group identity. Religion was seen as the means of moulding individual behaviour and the force responsible for shaping attitude to life and humanity in general. Few of the youth made an attempt to critically examine the beliefs of the religion; in the most part, there was an unquestioning belief in the religion and what it entailed for its followers. The combination of growing up in Muslim families and living in Jamia Nagar increased the inability to critique the religion. A common perception was that being Muslim in Jamia Nagar was the thread that unified residents. The celebrations during religious festivals such as Eid helped in asserting the feeling of brotherhood and the 'we'-feeling.

Being Muslim

When respondents were asked the defining features of what it meant to be a Muslim, the main responses related to their birth in a Muslim family, the rituals of Islam, religious education, religious markers and living in Jamia Nagar. Being born into a Muslim family reinforced the importance of religion. From childhood, the various rituals observed and followed by family members helped support the role and relevance of religion in their lives. This was the facet identified by most respondents. The various rituals associated with Islam were identified by respondents as defining to being Muslim. The role of the family, especially parents, was very important for learning the various rituals and practices. Practices such as the offering *namaz* (prayers) five times a day, observing *roza* during the holy month of Ramazan, offering *zakat*, going on the *haj* pilgrimage, reciting *kalma,* circumcision of male children and festivals such as Eid all contributed to the feeling of being Muslim. Though many respondents did not follow all customs, they felt a strong sense of connection with the practices, felt that it had an important ethical and moral base for all Muslims and this helped bring about a connection between Muslims from all parts of the world. It may be relevant to note that girls found the importance of rituals more than boys, as also identified by Mirza (1969).

More than half the respondents stated that *religious education* helped buttress the religious base of their identity. Being educated in schools in Jamia Nagar meant that they were given religious education in school and had to perforce read the *Quran* daily. This regular reading of the *Quran* was felt to be the means of learning right from wrong, and differentiating between good and bad, and leading a pure life with faith in the Almighty. This also implied that all respondents had a basic understanding of the tenets on which the religion was based, often in addition to the religious education being imparted at home. Religious education given at school served to buttress the foundations of the beliefs that had already been established at home. There was usually a consonance between value education given in the schools and the home, making the religious bases more relevant and meaningful.

Being Muslim was associated with certain *religious markers*, which had meaning only for the followers of Islam. The common religious markers mentioned by the respondents were *purdah* for women (which could vary in degrees from simple veiling to complete concealment), beard and cap for men and circumcision of male children. Religious markers were found to have distinct gender affiliations and expectations (the role and importance of *purdah* for the respondents is discussed in a later section). These religious markers helped separate the followers of Islam from others and were seen as instruments of identity reassertion, emphasising feelings of brotherhood within the community. It was also used to show the individual's love and dedication to God—the Almighty.

Living in an area like Jamia Nagar, which was a Muslim dominated enclave, served to reinforce the connection to Islam. Living in Jamia Nagar resembles "*clustering*" as defined by Latane and Bourgeois (2001), wherein residents share common attributes. The presence of numerous mosques, butchers selling *halal* meat, women in *purdah* and men with beards and caps all served to reinforce the importance of a cluster of individuals rather than one person or family alone. The degree of social influence that exists in closed enclaves such as Jamia Nagar creates expectations said the respondents, both men and women. There was more pressure on women to adhere to norms of dressing, interacting with members of the opposite sex and taking up family responsibilities. Some female respondents mentioned 'guilt' if they went against what was culturally and socially expected from them. This is consistent with Mirza (1969), who opined that women were more likely to be true to tradition, less likely to change and more likely to faithfully observe religious ritual often because of social and community pressure to conform. This further enhanced the role of women as guardians of the religion, carrying the faith forward. Living in Jamia Nagar meant that girls who decided to wear the *burqa* or *hijab* were looked at with approval and pride; it was an accepted norm for girls to make the transition from modest dressing to veil.

Religious Practices

Religious practices help in cementing the importance of religion in the life of an individual. As children grow in the family, there is a gradual socialization in the beliefs and practices that members of the family consider as important. Religious beliefs and practices are an important part of being Muslim. While 20 of the respondents said that they began practicing *namaz* and observing *roza* from childhood (between ages of 7-10 years), the other nine respondents said that they had begun to keep the necessary observances by the time they entered adolescence (between 11-15 years). It was mainly due to familial and parental influence that the religious practices began to gain importance in the lives of the respondents.

The family socialization into religion was gendered. Boys accompanied their fathers and older male members to the mosque for regular prayers. They were given the necessary education by their fathers on how to conduct themselves during the various rituals for festivals and also during mourning. The girls were given their religious education by their mothers at home. The close family involvement in the practice of the various rituals increases the collective identity associated with the religion, as mentioned by Chaudhary and Miller (2008) and Verkuyten (2005).

Though all the respondents mentioned the importance of *namaz* as a religious practice, only nine respondents (seven girls and two boys) said that they practiced *namaz* five times a day as required. Here again, it is observed that girls seem to be more particular in following the practices of the religion as compared to the men reinforcing. The other 20 respondents said that they had started following religious practice with great enthusiasm in childhood, but had slowly reduced the number of times they prayed daily to once or twice. The reasons for the change in practice was lack of time due to academic pressure, laziness and a lack of conviction in the need to pray five times a day in order to be called a Muslim. Though the practice of *namaz* five times a day was not followed by most respondents, the importance of Friday prayers was vital. None of the respondents could contemplate missing Friday prayers—even if the *namaz* on other days was perfunctory, the Friday prayers had to be observed. It may seem ironic, but even for those individuals who did not say *namaz* on a regular basis, the role of religion in shaping their identity was paramount.

Two male respondents, who said that praying and going to the mosque on Fridays was important, admitted that they themselves rarely went to the mosque and never prayed at home. These two respondents defined themselves as Muslims but did not follow the religious observances of prayer or even of observing *roza*.

Roza was another ritual practice that was voluntarily observed by 27 respondents who offered many reasons for following the practice. *Roza* was seen as a family affair, not just of kin but of all Muslims. During the month of Ramazan, the observance of *roza* helped draw together all Muslims, irrespective of caste or class, to come together and establish a sameness of routine from breaking the fast at dusk to partaking the feast at dawn. *Roza* was seen as a process of purification, with a scientific basis. The practice of *roza* helped keep evil thoughts at bay and brought observers closer to Allah. *Roza* was symbolic of self-control. Fasting was a process of controlling worldly pleasure and a means of showing their devotion to the Almighty.

Purdah and its Relevance

Purdah for Muslim women has been taken as an essential feature of Islam. In north India, *purdah* can be seen in Hindu communities as well, but the nature of the practice varies. The encompassing *burqa* for Muslim women is very distinct from the veil as used by Hindu women to ensure seclusion. The ritual nature of the veil is very different for both communities as well; for Hindus, the *purdah* restrictions are based on avoidance rules for male kin, while among Muslims, *purdah* restrictions apply only outside the immediate family unit (Papanek, 1973). Six of the female respondents who maintained the practice of *burqa* said that there had been no compulsion or force from the family to adopt the practice. In fact, one respondent said that she started the practice against the wishes of her father. For these six women, the *burqa* or *hijab* was a defining feature of their identity as mature Muslim women. In fact, four of these six women, said they would continue with the practice even at the cost of their career and irrespective of whether their future husbands or in-laws disapproved of the practice. These four girls were of the opinion that discontinuing the practice should be their decision alone. For all six girls who observed the *burqa,* it was the older women in the family who had been their role models.

These findings on *purdah* are similar to the findings of Vreede-De Steurs (1968) that Muslims in some parts of Delhi followed *purdah* strictly because being a minority group, it emphasized their religious heritage. Further, religious leaders have been active in emphasizing the importance of *purdah* for women. El Guindi (1999) observes that in parts of the Arab world, there has been a shift of young, educated women taking to the veil as an attempt to translate their ideals of Islam and serve as exemplars of Islamic practice. There is a reaffirming of Islamic ideals and morality accompanied by a rejection of the liberal practices of the majority group. Herman (2006) maintains that in India, attempts made by Muslims to establish their distinctiveness lead to the practice of using *purdah* to protect the identity of Islam.

Four of the six respondents who wore the *burqa* felt that they bore the brunt of suspicious looks from others around them—this suspicion came not just from people of other communities but also from other Muslims, especially those who did not observe the veil. As one respondent mentioned, Muslim girls who did not observe *hijab* perceived those who did as being backward and vulnerable and done due to compulsion and not out of choice.

The other five female respondents said that they had never felt the need to observe *purdah;* besides none of their immediate family members such as their mother or aunts followed the practice themselves. These women felt that observing *purdah* could come in the way of some career options and felt that they would prefer to keep their options open. These women were looking at careers where the *burqa* would be restrictive. These respondents felt that they should never be forced to wear the *burqa*. Even if their prospective in-laws wanted them to start wearing the *burqa,* they said they would not do so.

When male respondents were asked their views about the practice of *burqa* and *hijab,* with the exception of three respondents, all the others felt the *burqa* should be followed. They felt the *burqa* was beneficial for women and should be considered as such; they also felt the *burqa* was a protection against the odds and constraints of patriarchy and that the *burqa* freed a woman from the confines of her home and allowed her to move in society without fear of attracting the attention of men and fending off any unwelcome overtures. Wearing the *burqa* protected a woman's dignity; it prevented vulgarity by obscuring the contours of the body, thereby preventing sexual desires of men and subsequent attacks on women. Eight of the male respondents said that they would expect their prospective brides to wear the *burqa;* this was more in line with being the religious requirement as laid down by the *Quran*. Two of the male respondents said they did not allow their mother and sisters to leave home without wearing the *burqa* and would scold them if they did not observe *purdah* as required. Seven others felt that *purdah* was important, but they would leave the ultimate choice of observance to their mothers, sisters and prospective wives.

Three male respondents were totally against the practice of *purdah* and felt it was a reactionary practice, created for the purpose of suppressing women. They felt that the entire concept of veiling had been misinterpreted and that the *Quran* has never mentioned the need for *purdah* for women. Scholars such as El Guindi (1999) say that the Prophet was more concerned about men maintaining their bodily dignity and modesty during worship; and that there was negligible reference to a woman's dress.

Beard and Cap

Twenty-one respondents said that the beard was an important marker of the religion. The beard is considered as *"sunnah"* (practices followed by Prophet Muhammad) and men who kept a beard were considered to show their love and respect for the Prophet, who was identified as the ideal Muslim. The beard was considered as a sign of maturity for Muslim men; it set down certain rules that were expected to be followed—it should never be allowed to get unmanageable; on growing a beard, the man was supposed to respect all women, to be moderate in his behaviour, helpful to people around him. However, only five of the 18 respondents kept a beard and eight respondents said the beard had no Islamic relevance and was a mere fashion statement. The cap was considered to be a means of showing respect to God, by covering the head during *namaz*. Worn by male respondents who visited the mosque for prayers, the cap was considered an especially important identity marker for Muslim men during the month of Ramzan.

Interactions with the Majority Community

Of the 29 respondents in the study, only eight had studied in schools outside Jamia Nagar. These eight respondents said that their parents wanted them to have a better quality of education as compared to the education imparted by schools in the vicinity. Parents of these eight respondents were themselves educated and aware of opportunities available in schools outside Jamia Nagar. Of these eight respondents, three had gone on to pursue graduation in colleges outside Jamia Nagar. When asked about friends from other religious communities, most respondents said that they had acquaintances who were Hindus but their close friends were always Muslims. Two girls reported incidents of being treated badly by their Hindu friends, which led to avoid them.

All respondents felt that Muslims as a community were vulnerable and likely to be discriminated against by others around them. The perception was that Muslims were the first target for being picked-up and interrogated by the police for any suspected terrorist connections. While only two respondents in the study reported of any such incident happening to them, they were all able to relate an incident of discrimination that had occurred to someone they knew—a relative, friend or acquaintance. Two of the male respondents said that they had shaved off their beards after being stopped for minor traffic incidents and having to face questions about their religious affiliations. The respondents also felt that incidents like the demolition of the Babri Masjid and the Godhra massacre were responsible for the insecurity faced by Muslims, reinforcing the need for Muslims to live in closed enclaves such as Jamia Nagar for their personal safety and security. The respondents felt that as a group, Muslims were discriminated against, though they did not perceive the discrimination to be directed towards them as individuals. This finding was very similar to the findings of Taylor, Wright, Moghaddam and Lalonde (1990), who studied the discrepancy in discrimination between individuals and groups in Asian and Haitian women living in Canada. This difference between what happens at an individual level and at a group level could be due to an exaggeration of discrimination at a group level, which could very well be self-serving for the individual. There is also an additive phenomenon in this case, wherein the individual adds on experiences of discrimination of other people with whom the individual may be familiar with any personal experience of discrimination.

The perception of being the *"other"* was mentioned by some of the respondents. At least four respondents said that the majority community treated Muslims as 'foreigners' and were not seen as being an inseparable part of Indian society. All the respondents perceived themselves as being different from the majority community—a clear-cut 'us' *versus* 'them'—was articulated by the respondents.

It was felt that the media was responsible for the negative portrayal of Muslims in India. The perception among respondents was that the media projected a very derogatory and wrong impression about Islam. All Muslims were portrayed as fundamentalists and potentially linked to terrorist organisations. This, in turn, affected the way the general public views the Muslim.

Conclusion

The Muslim identity was the dominant identity for respondents in the present study. It appeared that the Muslim identity was more of an assigned or ascribed identity and not an asserted one. There seemed to be very little discussion or deliberation on the personal meaning that being a Muslim had for any of the youth. They were born Muslims and that was the identity they would carry with them. Being educated in minority institutions reinforced the feeling of being a Muslim, of being seen as the other. In this study, religion emerged as an essential category—all Muslims have the essential category of their religion to determine their identities. This essentialist understanding serves to maintain the existing social system, serving as justifications for the stereotypic beliefs and perpetuate the *status quo* (Verkuyten, 2005).

To explain the interactions between groups, it may be appropriate to try and analyse the situation using Berry's model (Berry, 2006) to primarily explain the acculturation of immigrant groups and understand the nature of interactions between minority and majority communities (Dovidio, Gaertner, Hodson, Houlette, and Johnson, 2005). In plural societies, when communities come into contact, certain questions arise about the nature of relations with the majority community, coupled with the desire to maintain one's own identity and culture. The greater the desire to maintain one's own identity and culture and minimise the interactions with the majority group, the greater the tendency for separation. Integration, on the other hand, comes up when there is a tendency to seek positive interactions with the majority community along with the desire to maintain one's own identity and culture.

It may be pertinent to re-examine the role played by the neighbourhood in which individuals reside. As mentioned earlier, the phenomenon of clustering occurs, while residing in places like Jamia Nagar. Furthermore, living in closed, sheltered enclaves further supports the tendency for separation (Phinney, Berry, Sam and Vedder, 2006). When living in closed communities, there are few opportunities for contact with peers from other communities; this means that people from other communities are perceived negatively and contact is avoided. While living in Jamia Nagar gives Muslims the impression of security and stability, it further reinforces the degrees of separation from the wider community. The opportunities for contact, either direct or extended, while living in Jamia Nagar are minimal. It is a case of residential segregation which reinforces other forms of segregation, such as distinct preferences for newspapers, shops and schools frequented.

A hypothesis put forth by Engineer (1991) says how communities which are socio-politically declining, reassert their religious beliefs and identity, giving itself an ideological edge over the other. This edge can serve as the means of drawing together all those who believe, giving the much-needed unity and solidarity necessary for group survival. The role of women in religious assertion has to be recognized as a powerful force of the assertion. The religious assertion seen in Jamia Nagar may be one form of the community coming together in the face of economic, social and political constraints faced by the group as a whole.

REFERENCES

Ahmad, A. 1993. *Muslims in India: Their Educational, Demographic and Socio-economic Status with Comparative Indicators for Hindus, Sikhs, Christians and other Communities, based on a Singular and Systematic Field Survey*. Vol. I. Delhi: Inter-India.

Berry, J.W. 2006. "The Contexts of Acculturation", in D.L. Sam & J.W. Berry (eds.), *The Cambridge Handbook of Acculturation Psychology*, pp. 27-42. Cambridge: Cambridge University Press.

Bhatia, S. 2007. *American Karma: Race, Culture and Identity in the Indian Diaspora*. New York: New York University Press.

Brewer, M.B. 2007. "The Social Psychology of Intergroup Relations: Social Categorization, In-group Bias and Out-group Prejudice", in A.W. Kruglanski & E.T. Higgins (eds.), *Social Psychology: Handbook of Basic Principles* (2nd edition), pp. 695-715. New York: The Guildford Press.

Cerulo, K. A. 1997. "Identity Construction: New Issues, New Directions", *Annual Review of Sociology*, 23: 385-409.

Chaudhury, S.R. and L. Miller. 2008. "Religious Identity Formation among Bangladeshi American Muslim Adolescents", *Journal of Adolescent Research*, 23(4): 383-410.

Dovidio, J.H., S.L. Gaertner, G. Hodson, M.A. Houlette, and K.M. Johnson. 2005. "Social Inclusion and Exclusion: Recategorization and the Perception of Intergroup Boundaries", in D. Abrams, M.A. Hogg & J.M. Marques (eds.), *The Social Psychology of Inclusion and Exclusion*, New York: Psychology Press, pp. 245-264.

Durkin, K. 2004. "Towards a Developmental Social Psychology of the Social Self", in Mark Bennett and Fabio Sami (eds.), *The Development of the Social Self*, New York: The Psychology Press, pp. 313-326.

El Guindi, F. 1999. *Veil: Modesty, Privacy and Resistance*, Oxford: Berg.

Engineer, Asghar Ali. 1991. "Remaking Indian Muslim Identity", *Economic and Political Weekly*, 26 (16): 1036-1038.

Erikson, E.H. 1966. "The Concept of Identity in Race Relations: Notes and Queries", *Daedalus*, 95 (1): 145–171.

Frable, D. E. S. 1997. "Gender, Racial, Ethnic, Sexual, and Class Identities", *Annual Review of Psychology*, 48: 139-162.

Gottschalk, P. 2005. *Beyond Hindu and Muslim: Multiple Identity in Narratives from Village India*, New York: Oxford University Press.

Herman, P. 2006. "India", in Kathryn, M. Coughlin (eds.), *Muslim Cultures Today: A Reference Guide*, Westport, CT: Greenwood Press, pp. 65-76.

Hewstone, M., E. Cairns, A. Voci, S. Paolini, F. McLernon, R.J. Crisp, U. Niens, and J. Craig. 2005. "Intergroup Contact in a Divided Society: Challenging Segregation in Northern Ireland", in D. Abrams, M.A. Hogg & J.M. Marques (eds.), *The Social Psychology of Inclusion and Exclusion*, pp. 265-292. New York: Psychology Press.

Howard, J.A. 2000. "Social Psychology of Identities", *Annual Review of Sociology*, 26: 367-393.

Ismail, S. 2004. "Being Muslim: Islam, Islamism and Identity Politics. The Politics of Identity-VI", in Richard Bellamy (ed.), *Government and Opposition*. 39(4): 614-631. Oxford: Blackwell Publishing.

Johnson, J.M. 2001. "In-Depth Interviewing", in J.F. Gubrium and J.A. Holstein (eds.), *Handbook of Interview Research: Context and Method*, Thousand Oaks, CA.: Sage, pp. 103-19.

Latane, B. and M.J. Bourgeois, 2001. "Dynamic Social Impact and the Consolidation, Clustering, Correlation and Continuing Diversity of Culture", in M.A. Hogg & R.S. Tindale (eds.), *Blackwell Handbook of Social Psychology: Group Processes*, pp. 235-258. Malden, MA: Blackwell Publishers.

Matsumoto, D. and L. Juang. 2003. *Culture and Psychology*. Belmont: Wadsworth Publishing.

Mirza, S.H. 1969. *Muslim Women's Role in the Pakistan Movement*, Lahore: Research Society of Pakistan, University of the Punjab.

Moghaddam, F.M. 2006. "Interobjectivity: The Collective Roots of Individual Consciousness and Social Identity", in Tom Postmes and Jolanda Jetten (eds.), *Individuality and Social Identity: Advances in Social Identity*, pp. 155-174. London: Sage Publications.

Papanek, H. 1973. "Purdah: Separate Worlds and Symbolic Shelter", *Comparative Studies in Society and History*, 15(3), pp. 289-325.

Peek, L. 2005. "Becoming Muslim: The Development of a Religious Identity", *Sociology of Religion*, 66(3), 215-242.

Phinney, J.S., J.W. Berry, P. Vedder, and K. Liebkind. 2006. "The Acculturation Experience: Attitudes, Identities and Behaviours of Immigrant Youth", in J.W. Berry, J.S. Phinney, D.L. Sam and P. Vedder (eds.), *Immigrant Youth in Cultural Transition: Acculturation, Identity and Adaptation Across National Contexts*, pp. 71-116. Marwah, New Jersey: Lawrence Earlbaum Associates, Inc.

Sachar Committee Report. 2006. *Social, Economic and Educational Status of Muslim Community of India*, Prime Minister's High Level Committee, Cabinet Secretariat, Government of India, New Delhi.

Seul, J.R. 1999. "Ours is the Way of God: Religion, Identity, and Intergroup Conflict". *Journal of Peace Research*, 36(5): 553-569.

Soofi, M.A. 2007. *Jamia Nagar: Delhi's Rich Muslim Ghetto*, 30 July. http://thedelhiwalla.blogspot.com/2007/07/jamia-nagar-delhis-rich-muslim-ghetto.html (accessed on 3rd October, 2009).

Soofi, M.A. 2008. *Dateline Jamia Nagar: No Pizzas for Muslims*, 5th April. http://thedelhiwalla.blogspot.com/2008/04/dateline-jamia-nagar-no-pizzas-for.html (accessed on 3rd October, 2009)

Stryker, S. 1980. *Symbolic Interactionism: A Social Structural Version*. Menlo Park, CA: Benjamin/Cumming.

Taylor, D.M., S.C. Wright, F. M. Moghaddam, and R.N. Lalonde. 1990. "The Personal/Group Discrimination Discrepancy: Perceiving my Group, but not Myself, to be a Target for Discrimination". *Personality and Social Psychology Bulletin*, 16(2): 254-262.

Verkuyten, M. 2005. *The Social Psychology of Ethnic Identity*. New York: Psychology Press.

Vreede-De Stuers, C. 1968. *Parda: A Study of Muslim Women's Life in Northern India*. New York: Humanities Press.

3

Muslims in Urban Ghettoes in Maharashtra: Life in Deprived Neighbourhoods

Abdul Haleem Kidwai

Although Muslims arrived at western coast of India as early as seventh century A.D., it was with the advent of the Alauddin Khilji in the early fourteenth century that the political presence of Muslims was established in the Deccan region. This was further consolidated by Muhammad bin Tughlaq's decision to transfer his capital from Delhi to Daulatabad. During the medieval period, there was considerable tussle for supremacy in the Marathi region of the Deccan. It was the battleground for the Bahmani Sultanates and its factions (Adil Shahis and Nizam Shahis), the Mughals and the Marathas and the Marathas emerged victorious. However, their dominance did not last long because the Marathas were defeated by the British and by early 19th century, it became a part of the British Empire and came to be known as the Bombay Presidency and Province.

Throughout the British period, Muslims in Mumbai had been underrepresented in the public services and in the education sector (Khalidi, 2005: 19-22).[1] As Christine Dobbin notes:

> "By the mid-nineteenth century, the Muslim community throughout the Presidency was noted for its unwillingness to espouse the new English education. The problem was particularly acute in the cities, where Muslims in general spoke Urdu and so required a separate system of elementary schools before they could go to even the Anglo-Vernacular level" (Dobbin, 1972: 229).

The present-day state of Maharashtra came into existence in 1960 and the Marathwada region, which was originally part of the Nizam-ruled state of Hyderabad, was included in Maharashtra. As per the Census of India 2001, Muslims constitute 10.6 per cent of the total population of Maharashtra and the rural-urban distribution of the community follows the national pattern, that is, they are more concentrated in urban areas (Government of India, 2006: 265–267).

Muslims and Urban Poverty

Maharashtra also follows another national pattern, that is, higher incidence of poverty among Muslims in urban centres. About 55 per cent of the urban Muslim population of Maharashtra is below poverty line while the figure for rural poverty is lower which stands at 28 per cent (Government of India, 2006: 367-368). This makes the study of urban poverty among Muslims in general and among Muslims of Maharashtra in particular important.

Although the Sachar Committee report has gathered a large body of data on the condition of Muslims in urban areas and it is established that they are the poorest community in urban areas (John and Mutatkar, 2005), there has been limited research on the *processes* responsible for their poverty. As Steve Wilkinson writes:

> "I argue that they (data in Sachar report) tell us far less than we might wish about the processes of discrimination and the major factors that may be responsible for Muslim disadvantage, which prevents us from designing the best government and NGO interventions to deal with problems the minority community faces" (Wilkinson, 2007: 832).

However, it is important to note that it is difficult with the available macro-level data to find out the underlying causes of poverty among Muslims in India. Therefore, in order to answer the question of *why* Muslims are poor, there is a need to gather data at the micro-level which will look into both the quantitative and qualitative aspects of poverty among Muslims. As Abusaleh Shariff argues, "There is a need to undertake focused micro-studies to clearly understand the nature, type and magnitude of socio-economic problems faced by various communities in rural and urban India" (Shariff, 1995: 2953). This chapter is an attempt to address the gap in the literature on poverty among Muslims by carrying out a micro-level study in Govandi, Mumbai, which is then supplemented with observations from Mumbra, Thane, a Muslim ghetto.

Muslims in Mumbai

As per Census of 1991, Muslims constituted about 17 per cent of Greater Mumbai's population and their presence in the city can be traced back to 14th century. Most of the Muslim population of Mumbai today are migrants from the Konkan region, Gujarat, Uttar Pradesh and Bihar. Although Muslims have always traditionally lived in their own community areas, they were also living in localities which had a diverse population. Furthermore, there is considerable diversity within the Muslim community of Mumbai. As Sameera Khan writes,

> "Historically, the Muslims of Mumbai have never been a single entity. Thus, the Muslim presence in Mumbai has always been marked by heterogeneity. Besides the general doctrinal classification of Shia and Sunni (which can be further divided by particular schools of theology, such as the Deobandis and Barelwis among the Sunnis), the city's Muslims can be categorised in several different ways including by place of origin, language, occupation, class and caste. The major groups in the city are the Dawoodi Bohras, the Sulaimani Bohras, the Aga Khani Khojas, the Halai Memons, the Kutchie Memons, the Konkani Muslims, the north Indian UP and Bihari Muslims, the Moplas, the Deccanis and the Iranis. In fact, by some estimates, Mumbai has the most heterogeneous grouping of Muslims than any other city in South Asia" (Khan, 2007: 1528).

However, with the riots of 1992-93, the cosmopolitan fabric of the city was torn apart. The Muslims were forced to flee 'unsafe' (read those areas where Muslims were few in numbers) to 'safe' areas, that is, Muslim majority areas. Govandi and Mumbra were such areas where Muslims fled to seek safety.

Govandi and Mumbra—A Brief Introduction

A report on "Socio-economic and Educational Status of Muslims in Mumbai and its Suburbs" (Jain and Shaban, 2009) describes Govandi as one of the poorest areas of Mumbai and has a high concentration of Muslim population. A sizeable section (41.3%) of Govandi's population lives below the poverty line and over four-fifth (81.4%) of the population lacks adequate housing and lives in slums/chawls. Govandi has a number of localities within it, the prominent ones being Rafiq Nagar, Tata Nagar, Lallubhai Compound, Gautam Nagar, Sanjay Nagar, Shivaji Nagar, Baiganwadi and Lotus Colony. Although a majority of the residents of Govandi share a common religious background, they come from a number of different states—Andhra Pradesh, Bihar, Gujarat, Karnataka, Maharashtra, Tamil Nadu, and Uttar Pradesh, among others. Out of this, the majority are from Uttar Pradesh (50.6%) followed by Maharashtra (25%) and Bihar (9.7%).

Mumbra comprises 19 wards and was brought under Thane Municipal Corporation in 1986. It is about 30 kilometres away from Govandi. It too, like Govandi, is a Muslim majority area and has a population of around 8 lakh.[2]

Although both these areas are marked by socio-economic deprivation (as will become evident in latter part of the chapter), in the public imagination these are conceived as 'criminal areas'. In the public discourse, the marginalization faced by the residents of these localities is pushed into the background. If one searches for Govandi or Mumbra in the websites of leading newspapers[3], the results are reports of theft, robbery, murder and other crimes whereas the poor socio-economic status of these localities remains underreported. Consequently, the attention of the public as well as the policy-makers shifts from tackling the multiple disadvantages faced by the community to restore 'law and order'. Therefore, the appropriate course of action is not to open more police stations but schools and hospitals. This will facilitate the inclusion of the community in the 'mainstream'.

Govandi and Mumbra: Facts and Perceptions

To understand the socio-economic situation, we collected data related to 100 Muslim households from two localities of Govandi, namely Rafi Nagar and Tata Nagar, in 2009. Out of these households, 51 were migrants from Uttar Pradesh and 49 from Bihar. The total population of these 100 households was 600, of which 321 belonged to migrant families from Uttar Pradesh and 279 were from Bihar. There were two main reasons for choosing migrants from these two states. The first was that these households are not only a religious minority but also a linguistic minority which in turn creates the potential for them to be doubly excluded—on the basis of religion as well as language. The second was that the author was fluent in both Hindi and Urdu which are the mother-tongues of these migrants, thus, making it easier to communicate and build rapport with the respondents. The population under study was mostly young. Only 6.7 per cent of the total population was above 60 years of age while 25 per cent of the population were children, that is, between the age group of 0 to 14 years. Thus, the significant share of the population was in the working age group, that is, between the age group of 15 to 60 years. The section below presents major findings from Govandi along with some details from Mumbra.

Living Conditions

Govandi is mostly a slum area and lacks adequate facilities. In Govandi, there are a few pockets of non-slum areas as well which have concrete buildings. However, most of these buildings are poorly maintained. In the sample covered for this study, 86 per cent of the households were living in slums or chawls. Even those residing in flats did not have spacious or luxurious flats. Most of these flats were small and cramped.

Since the houses were small, the room density in these areas was particularly high. About 69 per cent of the households had a room density ranging from 4 to 6 persons per room. Apart from the obvious issues of lack of privacy, and discomfort caused by these cramped conditions, there were certain hazards as well. For example, in a number of one room houses, the room was partitioned off by a curtain or a thin wooden plank with the living/sleeping area on the one side, and the kitchen on the other. Proper ventilation was missing in many cases which exposed the household members to smoke and heat. This was particularly dangerous for the toddlers and young children living in the house because they would play and move about near the kitchen rendering them vulnerable to getting burnt. Another issue that came to fore was that such an environment was not conducive to studying at all. One student remarked:

> "My family wants me to study and so do I, but it is simply not possible in my home. You can see yourself. There are 7 people living here (referring to his two room house) and someone or the other wants to watch TV or chit-chat. There is too much noise and distraction here. If the government cannot give us good homes it should at least provide libraries or reading rooms for us."

Another health hazard was the fact that these houses provided little or no protection from mosquitoes, leaving the residents prone to malaria, dengue and other diseases.

However, the most deplorable aspect is the lack of toilets and bathrooms within majority of houses. The people have to rely on community toilets and bathrooms. A number of households (29%) do not have a separate bathroom but the problem is more severe with regard to toilets within the houses (67% of the households do not have a toilet). The lack of cleanliness of community toilets possess serious health hazards and the dependence of the families on it also undermines their self-esteem. As one of the respondents said *"At times it is so humiliating to queue up for using the toilet… There is no 'izzat' (dignity) living like this."*

However, high rates of self-ownership of houses were reported. About 64 per cent of the households owned their house. The housing condition in Mumbra (of those covered in the sample) was significantly better than the residents of Govandi[4], in view of the fact that a large number of the respondents were living in flats and not in slums/chawls. Almost all households that the author covered had both bathroom and toilet inside. Most of the respondents in Mumbra had migrated from Mumbai (including from Govandi) after selling their property in Mumbai (which in most cases was a house in a slum/chawl). The money obtained from selling their property in Mumbai was enough to buy a flat in Mumbra. Some respondents also informed that they were left with enough money to expand their business even after purchasing a flat. However, a large number of residential buildings in Mumbra are 'unauthorized', that is, they have not been approved by the Thane Municipal Corporation (TMC). Hence, the tag of 'illegality' remains with a section of the population of both Govandi (due to non-authorisation of many slum pockets) and Mumbra.

One important aspect that emerged during the study was that the residents of both Govandi and Mumbra argued that they felt safe in these areas and that is one of the main reasons for them staying in there. In fact, even those respondents who could afford to move out of these areas said that they were unwilling to do so. They claimed that living in Muslim concentrated areas ensures the safety of life and limb which is not possible when living in a non-Muslim locality. A respondent from Govandi said:

> "We are doubly oppressed because we are poor and we are Muslims. Being poor means that the government doesn't care about us and being Muslim means that society thinks of us as outsiders. It is not uncommon to be labelled Pakistani or anti-Indian. At least living here (*in Govandi*) means that the community is with you. If I were to live somewhere else, who would be there to help me? Here, we have our own people who are always willing to assist."

A resident of Mumbra adds that "Living together gives us the chance to come together as a community and try to make improvements."

However, there is a darker side to this idea of 'living with the community'. It should not be perceived simplistically as a case of community cohesion or a 'preference' on the part of the Muslims to live together. A number of respondents claimed that communal politics and discrimination was responsible for their ghettoization. They argued that living together ensures physical safety and that the threat of violence had compelled them to migrate to these Muslim localities. This feeling was particularly strong in Mumbra because they were mostly victims of the 1992-93 riots in Mumbai. As one respondent puts it:

> "What is the point living anywhere else (that is anywhere away from the Muslims) when our lives can be shattered at any moment? The best example of it was Gujarat riots. What happened there? People who were very well established in business and professions were ruined

> overnight. And that is why Muslims have to stay together because if we are dispersed who will protect us? Even if we flourish outside but how long will that be for? A few years? A decade? A riot will happen and we will have to start from scratch. Why should we expose ourselves to such a risk?"

Apart from the security concern, there have been overt cases of discrimination in housing as well. During the interviews well-to-do professionals in Mumbra told us that Mumbra was not their first choice. They had looked for houses in other areas such as Airoli and Nerul but once it was known that they were Muslims, the owners refused to sell the flats to them. One respondent narrated his experience:

> "I found a flat in Nerul which was in my budget. It was a fairly new building and I was quite keen of getting it. When the time to finalize the deal came and the agent heard my name, he told me straightaway that no Muslims are allowed in this society. He wasn't even apologetic about it."

The experience of being denied housing on the basis of their Muslim identity was fairly common to the households interviewed; either they themselves had been denied or knew of someone first-hand who was refused because his/her being a Muslim. The respondents also said that some of the real-estate agents even justified this practice of refusing housing to Muslims. Some of the agents said that Muslims were considered dirty because they would slaughter animals and cook meat in the house, while other agents complained that since Muslims are connected with terrorists and criminals, it is not safe to provide them houses. However, the respondents pointed out that it was not the agents who were responsible but the property owners. In some cases, the agents had argued on behalf of their Muslim clients (that they should be allowed to rent/buy the flat) but the property owners refused.

Health

The issue of health is of major concern in both Govandi and Mumbra. Apart from the obvious bodily harm caused by illness, it is also responsible for a double economic loss if it befalls on the bread-winner. First, there is a loss of earnings caused by the inability to work. Second, the (meagre) savings are used up for treatments. This double-loss entailed as a consequence of poor health highlights the need and importance of affordable healthcare. The following words of a respondent make it very clear how difficult it is for the poor to cope with diseases:

> "Illness is a terrible thing. It should not happen to anyone and especially the poor. If we don't work, we will not have enough money for the treatment. On the other hand, if we do work while ill, the work will kill us. So, whether we work or not, it is death on either sides."

In Govandi, the people rely on a combination of both private and public health services but the primary sources of treatment for common illnesses are private practitioners. Of the total households covered, 71 per cent of them went to private practitioners while only 29 per cent went to government clinics. They gave several reasons for visiting a private practitioner instead of the government clinic. For example, the queues and waiting time at the government clinics was much more, in their opinion the private practitioners were more thorough in their diagnosis unlike the government clinics where it was a rushed affair, the private practitioners were more easily accessible and at the time convenient to the patient, etc. Another reason for preferring private healthcare was that some of the respondents chose to be treated by alternative medicine such as *hakimi, ayurveda,* etc. However, the number of households preferring alternative medicine was not substantial (only 21%). Of those who went to alternative doctors, homeopathy (11 per cent of the households) was slightly preferred over *hakims/ayurveds* (7%). There were mainly two reasons for choosing alternative medicine. The first, they felt that alternative medicines did not have side-effects unlike modern medicines, and second, these doctors practising alternative medicines charged less (both consulting fees and the cost of the medicines).

As far as the cases of surgeries and major illnesses are concerned, a overwhelmingly majority (91%) in Govandi go to government hospital while only a small section (9%) goes to private hospitals. The reason provided by the respondents for this was that the private hospitals were simply unaffordable for surgeries and major diseases.[5]

Although the huge chunk of the population goes to the local government hospital (called Shatabdi/Centenary Hospital), they are unhappy with it. It is ill-equipped to handle the number of cases that come to the hospital. It is a 210-bed hospital which serves a population of almost a million people (patients come from Govandi, Mankhurd, Trombay and Deonar). The Intensive Care Unit (ICU) has almost ceased to be an ICU and has become an extension of the general ward. Moreover, a number of machines in the ICU are not functional. The hospital also lacks a blood-bank. Furthermore, although the Brihan Mumbai Municipal Corporation (BMC) has sanctioned the budget[6] for renovating the hospital, no work till now has been carried out. Consequently, a number of emergency cases are referred to Sion hospital. This leaves the residents of Govandi (and adjoining areas) very vulnerable in case of accidents or other emergencies which require immediate attention.

The situation in Mumbra is even worse because there is no government hospital there. As a result people go to private hospital[7] or to go to Kalwa (a town nearby) or some other place out of Mumbra. A section also goes to Mumbai for medical treatment. And most of these respondents chose to go to public hospitals.

However, the more disturbing similarity between Govandi and Mumbra which emerged during the course of the study was the fact that the respondents complained that they had faced discrimination at the public hospitals (on the other hand, none of the respondents either from Govandi or Mumbra complained about being discriminated at private hospitals). A respondent from Govandi recalled the time when he was at the hospital because his wife was going to deliver his son. He recounted:

> "I was standing outside the room, when I overheard the midwife abusing my wife. She was saying that you Muslims keep producing children and all you bring are terrorists. I was shocked and deeply hurt by this."

However, not all the patients are passive victims of discrimination. For instance, one respondent who had faced a similar experience reported the nurse to the superintendent. He said:

> "I was furious when I heard the nurse shouting at my wife. So, I went to the superintendent and complained about it. I gave it in writing. I couldn't follow it because I didn't have the time. But I wanted to make a point that no one can do this. It is only when we (Muslims) assert our rights that will we be safe."

The complaint about being discriminated against was not limited to their ill-treatment at hospitals alone. They also complained that the failure to renovate the Shatabdi Hospital in Govandi and the absence of a government hospital in Mumbra was directly linked to the fact that most of the beneficiaries were Muslims. A resident of Govandi said:

> "The government has allocated the money but nothing has happened. I know there is corruption but at least some money is spent. In our case, they haven't spent a single rupee. Why is this? You can figure it out yourself."

The respondents in Mumbra took a similar position, as one respondent said:

> "Even places smaller than Mumbra have hospitals... look at Kalwa for example but why not for Mumbra? They (the Thane Municipal Corporation) are deliberately neglecting us. They know that the population of Mumbra is mostly Muslim, so they do not build hospitals."

Water Supply

The residents of Govandi struggle everyday to ensure that they collect enough water to meet their needs. They are not successful everyday. Not surprisingly, water is a precious commodity in Govandi and the residents have to pay an exorbitant amount for it. There is a 20-inch water main running along the Baji Prabhu Deshpande road which carries the bulk of the water to the slums. From this main pipe the BMC has laid a single official connection to the other side of the road for every lane from where there is a distributory network. However, there are a number of illegal connections, both overground and underground, directly from the main pipe. These illegal connections are under the control of the 'water-mafia'. They have installed booster pumps and draw the water out before it reaches its designated taps. So, it is only those who are living close to the main road are able to get water from the municipality. The rest have to purchase water or get it from the water tanker. The residents usually buy water in two forms.[8] One is in the form of a jerry-can and the other is to pay a monthly amount to the water-mafia to get water for a certain time from the tap (for example 20 minutes a day). The rate for a jerry-can of 40 litres of water is between Rs. 15 to Rs. 20 while the rate for drawing water from a tap is around Rs. 450–500 a month for 20 minutes a day.

Of the households covered, 64 per cent relied mainly on the informal water market to meet their needs while 34 per cent were dependent on the water tankers. Only 2 per cent of the households declared hand pump as the main source of water. The residents of Govandi have to buy water that too at such high rates from informal market, while BMC supplies water at only Rs. 2.25 per 1000 litres.[9]

Water is also an area of concern for the residents of Mumbra but the situation is not as bad as in Govandi. To begin with there is no 'water-mafia' (or at least in the form it exists in Govandi) and most of the people get their water supply from the municipality. The municipal water comes only for around one-and-a-half hours or two hours in a day which is inadequate for their needs. So, they supplement the Municipal water supply with bore-well water. Although the quality of the municipal water was satisfactory, sometimes the water-supply lines would get mixed with sewage water pipes causing muddy and dirty water to come through. In certain cases, water comes at very awkward times (around 4-5 a.m.) and in certain housing societies, the water supply is stopped on Fridays. Here again the water-cut was seen as a sign of discriminatory attitude of the water department. As one respondent argues:

> "How can the same water department supply us with water throughout the week except on Friday which is a holy day and we all bathe that day? They are well aware of what a water-cut on Friday entails and yet they indulge

in this act. It is a reminder by them that they are powerful while we are helpless. It is not only in the case of water that this happens. A few years ago, they cut the electricity on the evening of Eid but there was no power-cut during the Ganesha festival. Isn't this a clear case of discrimination?"

Sewerage System and Local Environment

The lack of an appropriate sewerage system and garbage disposal mechanism is a serious problem in both Govandi and Mumbra. Rafiq Nagar, Govandi, is built on a garbage dumping ground, so it has enormous environmental problem, air pollution, water pollution and dirt all-round. The condition of Mumbra is not too heartening either. Virtually every open space is a garbage dump and even roads are littered with rubbish. The people have complained that the Thane Municipal Corporation (TMC) does not perform its duties properly. What happens is that the TMC sweeper comes and makes heaps of the rubbish and leaves. Later on, the TMC garbage trucks arrive but the truck driver and his assistant refuse to put the garbage in the truck because they claim that it is the responsibility of the sweeper to load the garbage in the trucks. Consequently, the people have employed private sweepers to load the garbage in the TMC trucks. However, these private sweepers are usually employed by the housing societies and they are responsible for only their own particular housing society and at most the road outside. This still leaves the open areas neglected and there is no mechanism to clear away the trash which is dumped there. What makes the situation worse is that these are the same open areas where the children go out to play. When the locals were asked why they allowed the children to play in the filth, one of them responded:

> "They are kids and they feel like playing. We cannot keep them indoors all the time. And since there are no public parks or playgrounds for children to play, where else will they go?"

The sewage system is also in a very bad shape. In Govandi, there are 7 major *nallas* (drains) and they are in pathetic states.

Most of them are open and emanate a very foul smell. The stench is over-powering and unbearable in most cases. What makes the situation worse is the fact that almost all the drains are clogged with garbage and the sewage water hardly flows. These stagnant pools of water then become breeding grounds for mosquitoes. This problem is especially very severe during the monsoons and people fall prey to malaria, dengue and other dangerous diseases which can easily be prevented if a proper drainage is in place.

The poor sewage system and garbage disposal poses a health hazard. In fact, a doctor in Mumbra said:

> "People throw garbage everywhere and right out of their windows. They do not care about sanitation. This leads to a lot of health problems including jaundice, diarrhoea among others."

Socio-Economic Status of Muslims in Govandi and Mumbra

Employment and Level of Education

The employment status[10] of the Muslims in Govandi presents a disheartening figure on a number of counts. Very few of them were employed in the organized sector of the economy. Only 0.2 per cent of those in the working age group were employed in government service[11] and 2.8 per cent were working in the private sector. However, none of those working in the private sector had a permanent job; they were low level executives and had temporary contracts. In other words, 99.5 per cent of those employed did not have a secure job.

The overwhelming majority of the people were working in the unorganised sector.[12] Of those working, 43.5 per cent were self-employed, this was followed by those who were skilled labourers[13] (32.5%). However, it is important to bear in mind that these self-employed respondents are not high-end entrepreneurs but rather engaged in petty trades such as small shopkeeping, street vending, rag-picking etc. Similarly, those who were skilled did not possess skills which had high-returns but were mostly skilled in occupations which had low returns, i.e. plumbing, tailoring, etc. A clear gender division was evident in the employment status of Muslims in Govandi. Although the average work participation rate

(WPR)[14] was 47.3 per cent, it was heavily skewed towards men. The WPR for Muslim men was 77.6 per cent while for the Muslim women it was very low, that is, 15.5 per cent. These figures become starker when seen in absolute terms. Out of the 210 men in the age group of 15-60 years, there are 163 men who are working. On the other hand, out of the 200 women in the same age group only 31 are working.

Apart from the low WPR for women, they were concentrated in only a few occupations. In fact, there were no women who were in the government service and also there were no women who were skilled labourers.[15] Of those who were working, 48.3 per cent were home-based workers. Another important difference that emerged between men and women was in unemployment patterns.

The total percentage of men who were unemployed was 12.8 per cent while for women this figure was almost of half of the workforce (49.5%). However, the type of unemployment for men and women was different. Most of the men who were unemployed (85.1% of the unemployed) were those who were seeking work but could not find work while only 14.9 per cent of the unemployed men were not looking for work. Whereas for the women, the percentage of those who were unemployed and yet not looking for a job was quite high, both in absolute terms and relative terms (when compared with men). It stood at 53.5 per cent (while it was only 14.5% for men). What becomes evident from this is that the compulsion to look for work outside the home is much more for men than for women. As one respondent (a lady) said "It is the man's responsibility to earn a living but it is not so for a woman. The thing is that when a girl or woman is at home, she is not sitting idle. She keeps doing some work but when men are at home, they just laze around." This sentiment was echoed by a number of households. In other words, when a woman is not working, it is not perceived as unemployed but rather a conscious decision to stay at home. However, when a man is not working, it is seen as involuntary[16], that is, the man is just waiting for an opportunity to come by and he will start working. But this was not true for all households because 46.5 per cent of the unemployed women reported that they were seeking work but there was not available.

It was on the question of employment that the respondents spoke at length about how discriminated the Muslims were as compared to the rest of the population. The common complaint was that the government did not make any provisions for employing Muslims but affirmative action existed for other marginalized communities such as Scheduled Castes and Scheduled Tribes. One respondent said:

> "The condition of Muslims is as bad as Dalits, if not worse, but nothing is done for us. The government simply ignores the plight of Muslims. You have met so many people here already. Tell me how many have benefited from government jobs? Forget jobs, what about any government scheme? The fact is that the government is not interested in doing anything for the Muslims."

It was not only the government sector which was claimed to be discriminatory, many of the respondents said that the private sector was no better. They claimed that even private companies were not interested in hiring Muslims and especially residents of Govandi, who are perceived as anti-social elements and criminals. One respondent (whose son was a sales-agent for a private firm in Andheri) said:

> "Although my son is well-educated (he is a commerce graduate), he struggled to get a job. No one explicitly said that they refused his application because he was a Muslim and that too from Govandi, but I am sure that was the reason. In one place, where his application was rejected, a class 12th pass-out was selected. Then a friend of mine in my hometown (*a town in Bihar*) told me that his cousin's son had moved from Hyderabad to Mumbai and was working as a manager in a company in Andheri. I asked my son to get in touch with him and we were lucky because there were a few openings for salespersons in the company. Thanks to my friend's contact, my son got a job as a salesman. It would have been almost impossible to get this job had the manager been non-Muslim."

It indeed may be true that these Muslims might face discrimination during the selection process, especially for jobs in both the public and private sectors at the lower rung where the jobs are given at the discretion of the higher officials. However, during the study it emerged that very few people had actually applied for jobs.

Out of those in the age group of 15-60 years, 91.1 per cent had not ever applied for a job either in the private or public sector. Here again, a gender divide was evident. The number of men who had applied for a job was more than two-and-a-half times than women (12.8% men and 5% women). Another interesting fact which emerges when the data on job applications was disaggregated was that most of the applicants (73%) had applied for both private and government jobs whereas 27 per cent had applied only for private sector.

Two important reasons seem to be at play for the low-level of applications from the residents of Govandi. The first seems to be the low level of educational attainments amongst the Muslims of Govandi. The second is the widespread and strong perception that since they are Muslims they will not get the jobs, hence, leading to self-exclusion.

In Mumbra, a large number of the respondents were self-employed and were running small businesses. Here again a high concentration of Muslims in the unorganised sector was observed. One reason that came out very strongly for the prevalence of small businesses was found in the migration pattern of people to Mumbra. The population of Mumbra grew exponentially in the early years of 1990s after the Bombay riots. Muslims who had established businesses in Mumbai sold off all their property and migrated to Mumbra. When they came here they realized that the property was very cheap and that in the amount they had got from selling their houses, they could buy a house as well as have money left for investing in businesses/shops.

Conclusion

The following major conclusions can be drawn from this study. First, there is a strong sense of being discriminated against among the Muslims. Second, the macro-level findings of the Sachar Committee about the deprivation of Muslims are supported by this micro-level study. There is considerable lack of development in Muslim concentrated areas. The physical and social infrastructures remain in pathetic situations or non-existent. Even water and electricity is discriminately supplied to Muslim concentrated areas. Muslims suffer from a combination of backwardness, marginalization (both political and economic) and discrimination. These three factors—backwardness, marginalization and discrimination are deeply enmeshed together and it is difficult to isolate them from one another.

NOTES

1. In 1871, the share of Muslims in the population in Bombay (present Mumbai) was 15 per cent but there were only 9 per cent Muslims in the public services. By 1886, their share in the population had increased to 18.3 per cent but there were only 5.4 per cent Muslims in the Executive and Judicial Services (Khalidi, 2005: 19-22)
2. This information was obtained from the corporator's office of ward no. 113, Mumbra.
3. Hindustan Times, Times of India and DNA.
4. This is not to argue that the living conditions as a whole are better for the residents of Mumbra. The residents of Mumbra also suffer from the lack of adequate infrastructure; it is only that the housing conditions are better. In certain instances (e.g. roads), the conditions are much worse in Mumbra.
5. It is important to keep in mind that even for common illness, it is not easy for the poor to afford private doctors. The reason for visiting private practitioners is not so much as choice as is the fact that in the absence of well-equipped and effective public healthcare system, the respondents were forced to go to private practitioners at a very high cost.
6. Hindustan Times, *No Funds to Upgrade Govandi Hospital*, (March 20, 2010) (http://www.hindustantimes.com/News-Feed/mumbai/No-funds-to-upgrade-Govandi-hospital/Article1-521136.aspx Accessed on 28th November, 2010).
7. The name of the private hospital is Kalsekar. It is quite well-equipped but is not large enough to meet the needs of entire Mumbra. Furthermore, not everyone can afford the treatment there.
8. The author has labelled these two forms as the informal water market.
9. Rate listed in DNA (2010) *BMC's Outstanding Water Supply Dues*, (October 6, 2010) (http://www.dnaindia.com/mumbai/report_bmc-s-outstanding-water-supply-dues-rs3065-crore_1448272 Accessed on 9th November, 2010)
10. This has been gathered using Usual Principal Status (UPS). It has a reference period of one year and those who have worked for 6 months or more are considered as employed. Although the

employment pattern of those in the unorganized sector is very complex, for the purposes of this study, the worker has been assigned that employed category in which s/he claimed to have worked the most number of days during the period of his/her employment. For example, if a person has worked for 9 months in the year out of which s/he has worked for 6 months as a manual labourer and 3 months as a home-based worker, s/he shall be categorized as a manual labourer. Although this method is rather crude but it was the best that could be managed within the constraints of time and resources.

11. In absolute terms, there was employed only one person in the government. He was a lower-division clerk in the Municipal Corporation.
12. Although the precise definition of the unorganized sector remains contested, for the purposes of this study it has been defined as those enterprises whose activities or collection of data is not regulated under any legal provision or do not maintain any regular accounts and contractual labour agreements are absent.
13. These are skilled workers (electricians/plumbers/mechanics/carpenters/tailors and others) who are employed by someone else, usually in a shop owned by someone else and are paid every month. Those skilled workers who do not work for others are included in the self-employed category, though it is true that it is not possible to draw a sharp distinction between the skilled labourers, who are employees, and self-employed because the skilled employees also carry out extra independent work.
14. The WPR has been defined as the ratio between the number of workers in the age group 15-60 years and the total population of that age group.
15. One category of women who could have been included in skilled workers could have been women doing embroidery work. These women have been included in the home-based category because they work from home, usually they get the cloth and the design from outside and then they carry out the work at home.
16. Involuntary unemployed refers to those who are unemployed and are looking for work.

REFERENCES

Bharucha, R.N. 2006. *Yamuna Weeps Gently: A Journey into the Yamuna Pushta Slum Demolitions,* New Delhi: Sainathann Communication.

DNA. 2010. *BMC's Outstanding Water Supply Dues,* (October 6). *http://www.dnaindia.com/mumbai/report_bmc-s-outstanding-water-supply-dues-rs3065-crore_1448272* (accessed on 9th November, 2010).

Dobbin, Christine E. 1972. *Urban Leadership in Western India: Politics and Communities in Bombay City*. New Delhi: Oxford: Oxford University Press.

Government of India (Sachar Committee Report). 2006. "Social, Economic and Educational Status of the Muslim Community of India", Prime Minister's High Level Committee, Cabinet Secretariat, New Delhi.

Jain, Ranu and A. Shaban. 2009. "Socio-economic and Educational Status of Muslims in Mumbai and its Suburbs". Maharashtra State Minorities Commission, Government of Maharashtra, Mumbai.

John, Rijo and Rohit Mutatkar. 2005. "State-wise Estimates of Poverty among Religious Groups in India", *Economic and Political Weekly*, 40(13): 1337-1344.

Khalidi, Omar. 2005. *Muslims in Indian Economy*. New Delhi: Three Essays Collective.

Khan, Sameera. 2007. "Negotiating the Mohalla: Exclusion, Identity and Muslim Women in Mumbai", *Economic and Political Weekly*, 42(17): 1527-1533.

Shariff, Abusaleh. 1995. "Socio-Economic and Demographic Differentials between Hindus and Muslims in India", *Economic and Political Weekly*, 30(46): 2947-2953.

Wilkinson, Steven. 2007. "A Comment on the Analysis in Sachar Report", *Economic and Political Weekly*, 42(10): 832-836.

4

Journey towards Educational Equality: Enrolment Patterns of Muslim Girls in Mumbai and Surrounding Towns

Suniti Nagpurkar

"Ma'am, do you think 'like this (wearing burqa)' can I get a job in the media sector?"

"I just sent in my application without telling anyone at home and had decided that if I get an admission, I shall then somehow 'pattao' (get consent from) my family."

"Abhi yaha tak to aaye hai, shayad aage bhi kar payenge [*Now we have come at least this far, so may be in future also we may get (a chance) to do, what we wish to*].

[The above are some of the reactions of the Muslim girls, the respondents of this study, reflecting their apprehension, determination, acceptance, rejection, despair and hope.]

Background

The Sachar Committee Report (2006) laments the under-representation of Muslim community in education at all levels. And yet as per the report, the first striking feature is that the probability of completing different levels of school education has increased for all communities during 1983-2000. The sharpest rise has been in the probability of completing middle school for all communities including Muslims. Interestingly in the transition from the secondary to college education, Muslims perform somewhat better than Scheduled Castes (SCs) and Scheduled Tribes (STs), but lesser than other socio-religious categories. The Census of India 2001 has documented that the Muslim community has taken long strides and has caught up with the literacy levels, particularly female literacy in many States and Union Territories. It is observed that in 10 out of 21 major States, literacy levels of Muslims are higher than the State averages. In some of these States the literacy advantage in favour of Muslims is as high as 7-9 percentage point.

However, the Indian Muslims are not a homogeneous community (Engineer, 2000). They have sectarian, regional, caste and cultural differences which are quite crucial to the understanding of the developmental problems. Even in matters like literacy, family planning, social and economic development, there are significant regional and caste differences within the community. For instance it is observed that there is higher rate of literacy among the Kerala Muslims than Muslims in other regions. Even the rate of family planning among the Kerala Muslims is higher than the Muslims, say in Uttar Pradesh or Maharashtra. It is evident that rather than religion, it is the region that has an overbearing impact on the educational attainments of different communities (Hasan and Menon, 2005).

But since the turn of the century, situation is changing favourably. Muslims are paying more attention to education across the country. They have also become quite conscious of female education and now increasing number of Muslim women are taking to education. According to one survey (Hasan and Menon, 2004) in Uttar Pradesh Muslims, the percentage of Muslim girls appearing for SSC examination has increased. Not only this, but the success rate of Muslim girls also jumped nineteen times. It looks like Muslim girls are taking their studies much more seriously today than prior to 1990. This healthy trend seems to be persisting. Many Muslim girls are making it to the merit lists in Maharashtra also. According to Hasan and Menon (2004), the belief that Muslim parents are unwilling to send their daughters to a school is thing of the past and in fact though they themselves may not be educated, they want their daughters to attend schools and colleges and are thus willing to send them for not only primary or secondary levels, but even beyond.

Thus, while Muslims as a community, though a late starter in the field of education, is catching up. But the fate of Muslim women shows a large differential in their performance. Further, as is observed by some scholars, the conditions of Muslim women is never analysed beyond the cultural frame dominated by personal law and triple talaq. The forces like globalization, economic development, male migration, are very rarely factored in. One such study was by Aboobacker Siddique's (2005) detailed analysis of women representation in two districts—Muslim-majority Malappuram and Hindu dominated Kannur—in Malabar. It attempted to answer some questions: What environmental and socio-political factors make for Muslim women's impressive presence in Malappuram, but not in Kannur where Muslims constitute 27 per cent of population? His conclusion is that a combination of four factors has been enormously enabling: a sense of security, perceived backwardness, changing social mobility and phenomenal male migration to the Gulf.

Are then Muslim women less educated because of cultural compulsions or because they have not been provided with adequate educational facilities? Gender differentials in educational attainment and their links to other aspects of social attainment require particular attention. There is a need to explore whether and why development provides opportunity to certain groups of women to shed the excess patriarchal oppression; and not for certain other groups.

There is thus a need to investigate as to what is the connection between aspirations, rights and empowerment in different settings. Probably it is here that the role of Mumbai factor, its metropolitan, pluralistic nature can emerge strongly. With this background, the present chapter attempts to explore the following questions with regard to Muslims in Mumbai. What has been the attitude of Muslim girls and their parents towards higher education? Whether there is any definite trend with reference to the educational choices? What benefits do they expect from higher education? Has there been regional, social or any other dimension to it? What is the nature and extent of social interaction? What is the perception among Muslim women about the safety, liberating, and empowering social environment of Mumbai?

Mumbai, the most populated metropolitan city of India today, is unique in more than one way. It is very often called as 'mini

India'. The diversity of India in terms of its language, caste, class, creed and religion is fully represented in this city, in all its splendour, valour, dialectics and tensions. To a very great extent, Mumbai is the city of migrants. There have been inter-state and intra-state migrations to this city. The Muslims who have migrated to Mumbai constitute a heterogeneous community. They not only belong to different regions of India but also belong to different castes and class groups. The Marathi speaking community is the majority linguistic community in the city and is followed by Hindi, and Gujarati speakers, while the number of Urdu speakers as per 1991 and 2001census comes at fourth place. Between 1961 and 1991, Hindi speaking population in Mumbai increased from a mere 3,30,529 to 15,59,928, a whopping about 500 per cent increase. The number of Muslims in Mumbai was 16,70,170 in 1991 and 22,23,471 in 2001 (Census of India 1991, 2001). The population of the city of Mumbai as per Census 2001 was 119,78, 450 and Muslims constitute about 18.56 per cent of the total population.

Urdu speaking population is about 11,99,441. While some non-Muslims have Urdu as their mother tongue, their number may be very small. Thus, about 4,70,000 Muslims speak languages other than Urdu. To understand young Muslim girls' reality within Mumbai setting we have to bear all these differences in mind.

It is not the diversity but pluralistic nature of the society that results in the synergic benefits to the group. This indeed is the basis of the policy of affirmative action or positive discrimination being followed in many countries. One of the methods of assessing pluralistic policies and tendencies could be to study the public policy on education. In this context, it is worth noting that the Municipal Corporation of Greater Mumbai (MCGM) runs as many as 205 Urdu medium schools of primary and secondary levels in addition to about 1000 Marathi, Hindi, Gujarati, Kannada and Sindhi medium schools. At the higher education level, a glance at Departments of University of Mumbai in terms of subjects offered indicates that it is one of the most pluralistic universities of India today. This indicates that there is government initiative to support and maintain pluralistic culture. Thus, multi-lingual, multi-religious and migrants' city, Mumbai, provides a perfect setting of pluralistic culture. However, to assess the pluralist nature of the diversity it is necessary to assess the degree of assimilation, social interaction and exchange among diverse groups. If pluralism is served on the platter by the government it will have limited success, if any. In order to be successful it must also be on the agenda of the civil society.

The cosmopolitan pluralist trends are expected to bring about a good degree of exchange—cultural, behavioural, aspirational-among the population, particularly young population. Therefore, it was felt that the degree of freedom and empowerment experienced by Muslim girls in Mumbai is likely to be a result as much of their religio-cultural background as their exposure to a more liberal, free environment open to cross-country and international forces.

The University of Mumbai has seen a phenomenal enrolment of female students registered at various levels over a period of time. The enrolment of girls in various courses was 90,687 in 1997, and increased to 1,00,963 in 2001, and 1,14,606 in 2003. This size itself may create a strong demonstration effect, affecting the aspirational and behavioural pattern of young girls coming from diverse socio-cultural-economic background. Therefore, even for the community which is known more for restrictive, and rigid approach towards its women; the community where women are generally considered to be reeling under 'triple' deprivation arising due to patriarchy, gender and community forces, is likely to experience a major change in terms of its women's own self-empowerment, aspirations and future expectations etc. If this hypothesis is true then where would such a change manifest itself the most? It was thought that this change may be manifested most in terms of (a) presence of girls from this community in higher education, (b) work participation, (c) contemporary consumption patterns and lifestyle particularly the dress code, and (d) general sense of freedom of choice.

The present study was undertaken with reference to undergraduate general stream courses offered by University of Mumbai. It has two dimensions—quantitative and qualitative. Quantitative dimension is confined mainly to the enrolment data of the Institute of Distance Education (IDE), University of Mumbai that offers B.A. and B.Com. courses at undergraduate level along with other courses that are offered at the post graduate level. In addition to this, the enrolment data was taken from some of the sample colleges selected on the basis of location, co-education, and minority or non-minority status. The qualitative study was on the basis of focused group discussions and personal interviews. The reasons for selecting IDE were as follows:

(a) Availability of official data on enrolment over a period, from 1995-96 to 2006-07 which was expected to be useful to study the trend.

(b) Traditionally, Arts courses and off late Commerce courses are becoming very popular among girls going in for higher education. With a growing number of professional courses and colleges like Engineering, Pharmacy, Medicine, BMS, BMM, B.Sc. (IT), the general stream science courses have lost their appeal to students as is clear from the enrolment of students in science under various colleges of University of Mumbai. In all these colleges the science divisions, never get enough students and, therefore, are underutilized. Arts and Commerce have become preferred choices particularly for girls. It was thought that IDE that does not offer science courses may not leave a major lacuna in understanding the trend among Muslim girls in higher education. But still it is to be noted that this does leave out one major stream of higher education.

(c) The choice of IDE data was also on the assumption that the Muslim Community which is now emerging from being a late starter on the horizon of higher education to the stage where it registers the highest growth rate in female literacy, primary and secondary education, may still prefer distance education as a mode of receiving higher education for their girls rather than sending the girls to colleges. Therefore, enrolment data from distance education centre may provide a basis for capturing the emerging trends in Muslim girls' education.

Methodology

The data from the Institute of Distance Education (IDE), University of Mumbai, was collected from the annual enrolment data maintained by the institute itself. However, the institute maintains the data classified on the basis of faculty and gender only. Therefore, from the annual enrolment data the researcher separated the data on Muslim girls on the basis of names, surnames, father's name and mother's name. The data relates to the period from 1995 to 2006.

In addition to the data on enrolment with IDE, some sample data was collected from some of the colleges so as to have a better understanding about the enrolment trend among Muslim girls in higher education. The colleges were selected from the main city as well as suburbs. Care was taken to cover the colleges run by Muslim as well as non-Muslim educational trusts. The colleges were selected on the following basis: (a) two colleges run by Muslim education trust—one each from eastern and central suburbs of Mumbai; (b) four non-Muslim trust colleges were selected— two from the main city, run by Catholic Educational trusts, and two from suburbs run by Sindhi and Gujarati Society Educational trusts; (c) to have a more meaningful insight into the Mumbai factor with reference to the enrolment, a comparative view and data was sought. For this a *mofussil* area college from Konkan region that comes under Mumbai University was selected. Konkan region comprises Thane, Raigad, Ratnagiri and Sindhudurg districts. Konkan region has a sizable amount of Muslim population. Further, it also has a large galaxy of social thinkers and educationists; the most prominent among them is Bharat Ratna D.K. Karve, a proponent of the Women's education in Maharashtra. Dapoli is a taluka in Ratnagiri District. It has an Agricultural University, a college of Homeopathy and two colleges of Arts, Science and Commerce. Out of these, Belose College of Arts and Commerce was selected. In all, 21 Muslim girls currently studying in graduation in first year (F.Y.) and third year (T.Y.) Arts and Commerce faculties were interviewed and the group discussion was held. The college authorities were very co-operative and willingly and promptly provided the database as well as arranged the interaction with Muslim girl students.

The Enrolment at IDE

The enrolment data from IDE relates to the period 1995-96 to 2005-06. The IDE data revealed that over this period the number of Muslim girls enrolled at first year in Arts and Commerce stream has increased from 128 to 538 a little over fourfold increase in a decade's time as is seen in Table 4.1. During this period, the percentage of Muslim girls to total open category girls registers the growth from 4.71 per cent to 9.52 per cent (Table 4.2).

Table 4.1: Muslim Girls' Enrolment at Institute of Distance Education at First Year (F.Y.) B.A. and (F.Y.) B.Com.

Year	No. of Muslim Girls	Per cent Growth Rate in Enrolment
1995-96	128	–
1996-97	140	9.37
1997-98	173	23.57
1998-99	307	77.45
1999-2000	265	–13.68
2000-01	433	63.40
2001-02	317	–26.79
2002-03	506	59.62
2003-04	440	–13.04
2004-05	504	14.54
2005-06	538	6.74

Source: Data collected from IDE, Mumbai University, Mumbai.

There are of course some year on year variations which cannot be fully explained (Table 4.1). As can be observed from the year 1997-98 to 2003-04, the trend has been alternating large ups and downs. The plausible explanatory factors may be:

(1) *Higher secondary results particularly of Urdu medium higher secondary section*. Since a large number of students taking admission to distance education come from vernacular medium. As is observed that in earlier period, the results of Urdu medium schools used to be very low but of late there has been a considerable improvement in their results. Till a few years ago, the pass percentage of Urdu medium schools would be around 50 per cent which have now increased beyond 70 per cent. There is a positive change in the performance of Urdu medium schools.

(2) *The change in higher secondary syllabus*. For instance, 1997-98 was the last year of the then syllabus of H.S.C. Board. Generally, at such turning points there are noticeable differences in H.S.C. results. In the subsequent year, therefore, the F.Y. admissions are likely to be more. That is why overall girls' enrolment in 1998-99 has also increased by a large number.

(3) The University of Mumbai has introduced a number of undergraduate courses in B.A., B.Sc., and B.Com. as self-financing courses. In 1999, it introduced B.M.S. in

Commerce. In 2000-01, it introduced B.M.M. in Arts. In 2003, it introduced B.Com. (Banking Insurance) and in 2005-06 it introduced B.Com. (Accounting and Finance). This also may have some bearing on the enrolment at Distance education though indirectly.

Table 4.2: Enrolment of Girls in Open Category at (F.Y.) B.A. & (F.Y.) B.Com. and the share of Muslim Girls at IDE

Year	*Total Enrolment of Girls in Open Category*	*Enrolment of Muslim Girls*
1995-96	1939	128 (6.60%)
1996-97	2386	140 (5.86%)
1997-98	2613	173 (6.62%)
1998-99	3947	307 (7.77%)
1999-2000	3189	265 (8.30%)
2000-01	5243	433 (8.25%)
2001-02	6728	317 (4.71%)
2002-03	5315	506 (9.52%)
2003-04	5440	440 (8.08%)
2004-05	6582	504 (7.65%)
2005-06	6041	538 (8.90%)

Source: Data collected from IDE, Mumbai University, Mumbai.

Table 4.2 shows that the enrolment of Muslim girls in open category at IDE has improved over the years. The percentage of enrolment of Muslim girls in relation to that of non-Muslim girls in Arts faculty has been around 9 per cent while in Commerce, it has been about 5.6 per cent. From the data in Table 4.3, it seems that Muslim girls show more preference to Arts than Commerce, though since 2000-01 there is an improvement in the absolute number of Muslim girls joining Commerce stream. For instance, in 1999-2000 there were only 75 Muslim girls who took admission in Commerce at IDE, in 2004-05 the number increased to 232. Between 1995-96 and 1999-2000, the enrolment of Muslim girls under Commerce stream at IDE increased only to 75 from 61, while in the next 5 years it increased from 75 to 232.

The enrolment data collected from the IDE and some of the city colleges also indicate a rise in the absolute number of Muslim girls in higher education. Their choice of faculty, off late has shown that some of them are taking up professional courses like B.M.S., B.M.M. etc. But the percentage share of Muslim girls relative to the total girls' enrolment is still small. There is, however, a considerable amount of jump in the rate of growth in the past few years.

Table 4.3: Faculty-wise Enrolment of Girls at the Institute of Distance Education, Mumbai

Year	*Enrolment of Girls in F.Y. B.A.*		*Enrolment of Girls in F.Y. B.Com.*	
	Open Category	*Muslim Girls*	*Open Category*	*Muslim Girls*
1995-96	758	67 (8.83%)	1181	61 (5.16%)
1996-97	1201	70 (5.82%)	1185	70 (5.90%)
1997-98	1284	110 (8.56%)	1329	63 (4.74%)
1998-99	2021	204 (10.09%)	1926	103 (5.34%)
1999-2000	1744	190 (10.89%)	1445	75 (5.19%)
2000-01	2413	250 (10.36%)	2830	183 (6.46%)
2001-02	2845	206 (7.24%)	3883	111 (2.85%)
2002-03	2339	301 (12.89%)	2976	200 (6.72%)
2003-04	2349	199 (8.47%)	3091	241(7.79%)
2004-05	2945	272 (9.23%)	3637	232 (6.37%)

Source: Data collected from IDE, Mumbai University, Mumbai.

With reference to the student enrolment, it is to be noted that the Muslims often come under open category. While a good amount of facilitation is being offered by starting a large number of Urdu medium schools (by some informal sources, it is the only section which is growing as far as non-English schools are concerned) there also is an initiative that is being shown by the Muslim community philanthropic community or social groups to expand the education among the Muslim community which is considered as socially and educationally backward or less privileged.

In Maharashtra, a desirable shift in favour of education is seen in the last few years among Muslims. In fact in early 1990s, a large number of Urdu medium schools used to be almost 'zero per cent' results schools. Thus, the feeder group itself used to be small. The gross enrolment rate used to get adversely affected while effective enrolment rate may have been slightly better. This suggests that the concentrated efforts need to be made both by public as well as philanthropic bodies to improve enrolment, participation and successful completion of middle and higher secondary levels of education.

To understand the faces behind figures it was necessary to interact with the Muslim girls in higher education. The author interacted with 101 Muslim girls studying in different colleges in different courses at undergraduate level. Table 4.4 shows the faculty-wise distribution of respondents.

Table 4.4: Faculty-wise Distribution of the Respondents

Programme	*Respondents*
B.A.	52
B.Com.	24
B.Sc.	8
BMM	5
BMS	6
B.Sc.IT	4
Engineering	2
Total	101

Source: Author's interaction with Muslim girls.

The respondents came from different subsets. Out of 101 girls, 7 were from Bohra community, 2 from Khoja community, 3 from Sunni Deobandis, 3 Sunnis but Ansaris, and the rest called themselves Sunnis without identifying with any specific sects, like Deobandi or Brelvi.

Family Background of the Students

To assess their economic conditions, the students were asked to indicate the family possessions in terms of consumer durables. The interview schedule contained a question "What does your family possess from the following list?" And a list of assets like house, fridge, T.V. computer, car, telephone, mobile phone was provided. Most of them, reportedly have owned house, fridge, and T.V. and phone. Fathers' occupation mostly showed business, self-employment. About 78 girls reported their father's occupation as business. The businesses varied from construction to garments, textiles to electrical shops, garage to electronic, other small shops to fruit vending, tailoring, etc. Some reported their father as *jari* workers or *karigar* (artisan). Only 15 respondents said that their fathers are working with private companies or public sector companies. While 10 respondents said that their fathers were working abroad, but 5 had their brothers or uncles working abroad who were providing financial assistance to them.

Educational Background of the Respondents' Families

A very large number of girls came from the families where parents were not highly educated. The fathers usually had studied up to 10th or 12th standard and the mothers were just literates. Among the respondents who were currently studying under professional self-financing courses, only four respondents had their fathers working in some companies and were highly educated, one was a postgraduate, two engineers, and three graduates in Arts. Four respondents' mothers had done either graduation or post-graduation. Mothers of two respondents were working, both teachers. The students were also asked about the educational levels in their extended families. They were asked to indicate the highest level of education so far attained in the family. It was observed that from the entire group, very few respondents actually came from well-educated family backgrounds. The percentage of respondents, who can be said to have a well-educated family background, where both the parents were at least graduates, was 6 per cent.

Table 4.5: Distribution of Respondents by the Education Background of the Parents

Parents	*Level of Education*			
	Postgraduate	*Above HSC but less than Postgraduate*	*Below HSC*	*Illiterate*
Father	02	26	65	08
Mother	nil	05	85	10

Source: Compiled by the author.

Table 4.5 shows the educational background of the respondents' families. As can be seen from the table, the largest concentration in both the groups is in the category where the parents have not completed even higher secondary level of schooling. Among this category, a large number was even below SSC level. The respondents were also asked to indicate the highest educational qualifications in the extended families. Very few of them actually reported a very highly educated extended family background. Further, it was observed that many of the extended families are still in their native places and it is the cosmopolitan urbanized environment that has greatly influenced the educational attainment of the families who have shifted to Mumbai. In most of the responses, the extended family level of education was lower than that of their own families. Similarly, the educational levels of cousins of the respondents who have stayed back in their native place were usually lower than the respondents in majority of the cases.

Lack of higher education on the part of parents means that they themselves are unaware of new avenues of education and their usefulness. Their social exposure also may not provide information on newer options. Further, while their economic conditions may not be very pressing the social backwardness may result in continuation of traditional orthodox attitudes towards women's role, education and labour-force participation. As the result of which while it is true that there is considerably lesser amount of parental opposition to girls' higher education, they may still prefer general Arts, Commerce and Science courses.

Respondents' Perception about views of Islam on Higher Education

With regard to Muslim girls coming to higher education, all the respondents said that Quran does encourage for education. Many quoted prophet saying that "one can even go to China if necessary for education". The girls promptly explained that during prophet's time China was a distant place, very far off, but still prophet encouraged to go anywhere in the world for education and this freedom was available to all, men as well as women.

On this background, the reference was made to one newspaper report citing the case of two Muslim sisters going to city colleges like Xavier's and Elphinstone threatened, and assaulted for attending college by a neighbour who considered it anti-Islamic. Surprisingly not all had heard or read about it. This reflected the fact that very few read newspapers regularly, even national daily like *Times of India* where the report had appeared on the first page.

Incidentally, many said they do not get the newspaper regularly at home. Some said that they get Urdu daily which they read once in a while and others who get the newspaper at home do not read it on regular basis. The college libraries are visited rarely and are visited only for study books, note completions etc. Many students said that they have household chores like cooking to be done or classes to attend or have no permission to stay back in the college after the lectures with the result that they hardly visit library.

On soliciting their views on the incident that was reported in the paper, they all said that this only reflects the uneducated minds and wrong interpretation of Quran. In fact, many opined that it is wrong interpretation of Quran by some uneducated people or people with vested interests that has given Islam a wrong image and one of the things they would like to sea changed is people's mindset. They were hopeful that with spread of education, things will improve greatly.

On Government Policy and Support

During the discussion on joining higher education, question invariably came up on the policy of reservations and quota. Most were against any reservation or quota on the basis of caste (among Muslims) or religion, though they were of the opinion that some consideration for economically backward students is necessary. So poverty rather than the caste or religion should be the basis of reservation for Muslims.

While nobody mentioned it explicitly, it was observed that in this case more important facilitating factor has been the good number of Urdu medium schools started by the government that has helped. Among the respondents 23 per cent came from English medium schools, about 5 per cent came from Marathi and 7 per cent came from Hindi and Gujarati medium schools, while almost 65 per cent had studied in Urdu medium up to SSC. Some of them studied in Urdu medium up to HSC. Availability of Urdu medium schools seems to have helped at the initial stages of education. Some of the respondents also mentioned that after HSC many of their cousins and friends join D.Ed. since that helps them to get a job quickly. However, there is also an increasing realization about the importance of English medium education. Some respondents explicitly mentioned that there is a need for Muslim community to realize and accept this, if they really want to stand the increasing competition.

Parental Support

While most students except from Sophia and Xavier College were first time college goers, getting the permission was not a major problem in most cases. Of course, there were some instances where some sort of cajoling was required to be done. In such cases, the girls navigated and negotiated their entry into colleges by adopting various methods. Some agreed to go to the colleges and courses, both selected by parents. Two respondents from Somaiya College for instance said that they wanted to go to little far away college and wanted to do commerce but parents put the condition of joining Somaiya College only since it was about 5 to 10 minutes distance from where they live and join Arts since "there are no boys; at least not many" in Arts and "as it is, doing Commerce is not a great advantage because you are not going to take up any job". Some others reported that "they were given the permission to join college on the conditions that they would wear *Burqa*, not to talk or mix with boys".

Some girls had to seek the permission by carrying out the "war of nerves". Of course, here the problems were not that they were not allowed to join the college as such. The differences were in terms of which college and what course. Gulshan, a student from Akbar Peerbhoy College, for instance, wanted to pursue her Bachelor degree in Mass Media and make a career in Media. Her father wanted her to pursue Medicine. To fulfil her father's wish she gave it a try and joined science for higher secondary, got good marks in H.S.C. exam but felt that she would not be able to take up so much strain and in any case she herself did not want to join medicine. She was fascinated by Media and wanted to do BMM (Bachelor of Mass Media) which her family was not in favour of. She wears a head scarf and an Irani Coat but not *burqa*. Her father tried to talk her out off her media craze "With attire like this who will give you a job in Media?" Further media field is not considered respectable by many families/community. Gulshan, however, stood her ground and stopped talking to anybody in the house and finally obtained the permission.

Gulshan comes from a very well-to-do and educated family. Her father is a graduate and is employed with Bombay Port Trust. Her mother has studied up to 10th class and Gulshan is the only child. The family consists of grandparents, parents and her. The family finally had to give in to the daughter's wish and she was allowed to join BMM course on the condition that she should not change her dress code, must always wear a head scarf and the Irani coat. This was alright with her. A very confident, friendly, bubbly and warm person she is currently studying in third year advertising in Akbar Peerbhoy College and wonders whether "(she will) get a job with a dress like this?"

According to some professors and authorities at Burhani College, two media components in the BMM course, namely film and TV are the kind of deterrents since many of the Muslim families do not consider these sectors respectable. In fact, in their college initially no Muslim girl enrolled for this course though now slowly and gradually some of them are coming forward.

The story of two other girls, both sisters doing their English literature major at Somaiya College is more or less similar. Their father is a mechanical engineer. Mother has done graduation in Urdu literature. The parents wanted them to pursue medicine. But they wanted "to do Arts, particularly literature, marry rich persons and settle". Initially parents were disappointed but eventually gave them the permission to join Arts and major in literature.

The families where both the parents had low education, 20 per cent of girls reported that their fathers want them to complete their studies, and equip and empower themselves to be able to stand on their feet if any such need ever arise in future.

About 10 per cent said that their mothers, though less educated or completely uneducated wanted the daughters to complete higher studies in spite of the fathers having some resistance or opposition for their higher studies. Some stated that it is their mothers who insisted on sending these girls to English medium schools, though the mothers themselves were not much educated. In a few cases, the elder brothers who themselves had studied up to 12th only, were encouraging and financially supportive.

Many of them said that they are much better off than their own sisters who were married off after completing middle or higher education as well as better off than their cousins who were denied this opportunity. About 70 per cent of the respondents came from the families where their cousins in their own age-group were denied this opportunity.

During the discussion it emerged that despite the community opposition and reservations at times, the parents, though less educated themselves, have agreed sometimes somewhat against their own wish or fear of social censure, to send their daughters to higher education and the daughters seem to realize this wanting to their best for themselves, for their family and also for the community.

About 12 per cent of the students reported that in spite of opposition from home they have persisted and have joined colleges. Some of the girls even reported initial non-co-operation from mothers who gave lot of household chores to them making it very difficult for them to find time for studies. Two of the respondents have decided upon the strategy of negotiating daily time-table where 1 or 2 hours are given for household chores but not beyond that. One of them said that she has decided for the time being not to participate in the household chores. She feels bad that the mother has to do all the works for the family of nine members as family couldn't afford any paid help—the father is the *jari* worker—but she is hopeful that once she completes the education, she will be able to help the situation and help improve the financial conditions of the family.

Household chore is one major issue. Many reported that they have to cook, clean and wash clothes everyday and it was the condition put to them before giving them the permission. Some of the respondents have taken up entire responsibility of cooking etc. for Sundays. It was reported by most that the girls must not only be well trained in cooking and all the household chores but should be doing it regularly so as to remain 'used to' household chores no matter what education they take or whether they work for wage employment or not, later in life. In other words, the traditional family gender roles continue to dominate the expectations of parents with reference to their daughters though they have conceded to their demands of higher education.

Other Support Systems

Given this social background, who prompted them to take up not only higher education but in some cases professional education? As is said, behind every successful man there is a woman, we can say "behind every such Muslim girl in higher education, there is a teacher telling her to move ahead." Many of the respondents categorically mentioned the names of teachers who not only showed them the way but also persuaded their parents to allow the daughters to join the course of their choice. Rehana for instance who is currently doing B.Sc. Physics gratefully acknowledges Mr. Hassan, her teacher, who somehow talked to her father to allow her to join college. Similarly, Farah studying at Sophiya College, mentions her teacher Mr. Hussain as her role model and says because of him she is in the college. In fact, she had taken admission to First Year (FY) at Sophiya College last year, still her parents particularly at the instance of her sister's husband did not let her attend the college with the result that the admission lapsed. Farah's father has a workshop where he employs 6-7 workers making suitcases. So economically they were better off; still the parents would not pay for her college fees. She therefore started giving tuitions and collected enough money to pay for the fees and yet her admission lapsed as she was not allowed to attend the college. It is her persistent efforts against all odds, along with her teachers' persuasion of her parents that she has again taken the admission to F.Y. Science at Sophiya College. Similarly, Fatima, the eldest among seven siblings with very low educational background of parents, is studying in B.Sc. IT, thanks to her teachers. Besides such specific instances many respondents gratefully remembered their school and Jr. College teachers who played a very important role in making it possible for them to attend college.

About 10 per cent of the respondents expressed a desire to start library and social awareness forum for the girls from their own community who were not so fortunate to get opportunity. About 14 per cent said that some learned people from their own community should conduct camps for Muslim girls and women to discuss the preaching of Quran, the rights that Quran has given to them and many other social issues. They themselves have reportedly attended such workshops and have benefited from them.

About 7 per cent are already actively involved with some social support groups, well organized forums like Aga Khan Foundation. This foundation is working hard to spread the awareness among the Muslim community and trying to bring them in the mainstream by encouraging families to send their children to schools and colleges. The foundation also provides financial support to the needy people in the community, bearing the cost of the education for the children. The respondents said that they are associated with the wing that is working to reduce the drop-out rate among Muslims.

On Education and Career Choices

It is usually believed that the courses with "premium" are usually beyond the reach of the Muslim students, particularly Muslim girls. Since the turn of the century, University of Mumbai has introduced a battery of self-financing professional courses like B.M.S., B.Com. with Banking and Insurance, B.Com. with Accounting and Finance B.M.M., B.Sc. IT, B.Sc. Computers, etc. In this sample selected randomly, out of 101 respondents 17 students were from these courses. Of course, since the author visited self-financing courses section of Akbar Peerbhoy College of Commerce and Economics (the college is managed by Anjuman-i-Islam), it was but natural that their number should appear so large in the sample of this study.

But the colleges where the author visited both sections, the presence of Muslim girls in these courses was very marginal. In Somaiya College for instance in the B.Com. (Accounting and Finance), there was not a single Muslim girl, though out of the total strength of 75 students there were 30 girls. Same was the case with their B.M.S. section. In 2005-06, there were only 2 Muslim girls in this section which had the strength of 120 students out of which about 50 per cent were girls. Burhani College introduced the BMS, BMM courses from 2003-04. However, for the first two years not a single girl took admission to these courses, though in their regular Arts and Commerce stream, a large number of Muslim girls had enrolled as is seen from Table 4.6. In 2005-06, there were 3 girls in a class of 60, out of which only 1 was Muslim. Tables 4.6 and 4.7 throw light on this.

What is significant is a sudden jump in the last two years. The change seems to have become steady. Similar is the experience of Akbar Peerbhoy College. Thus, while it is true that presently the self-financing courses are not very popular with Muslim girls, gradually they are turning towards it. Similar is the trend observed at V.E.S. College of Arts, Science, and Commerce, Chembur, Mumbai, as may be seen from Tables 4.8 and 4.9.

Table 4.6: Faculty-wise Total Enrolment of Girls and Number of Muslim Girls in Burhani College (a Minority College) in Mumbai

Year	*Arts*		*Commerce*	
	Total	*Muslim*	*Total*	*Muslim*
2003-04	127	125	353	288
2004-05	144	137	286	252
2005-06	146	141	260	223
2006-07	119	116	248	218

Source: Office of the College.

Table 4.7: Enrolment of Muslim Girls in the Self-Financing Courses at F.Y. Level in Burhani College (a Minority College) in Mumbai

Year	*BMS*		*BMM*	
	Size of the Class	*No. of Muslim Girls*	*Size of the Class*	*No. of Muslim Girls*
2003-04	60	Nil	15	Nil
2004-05	60	Nil	17	Nil
2005-06	60	01	17	Nil
2006-07	60	11	23	04
2007-08	120	14	37	12

Source: Office of the College.

Table 4.8: Enrolment of Muslim Girls in the Self-financing Courses at Non-Muslim Educational Trust College (V.E.S. College of Arts, Science, and Commerce, Chembur) at F.Y. level

Year	*BMS*		*BMM*		*B.Sc. IT*	
	Size of the Class	*No. of Muslim Girls*	*Size of the Class*	*No. of Muslim Girls*	*Size of the Class*	*No. of Muslim Girls*
2003-04	60	Nil	50	Nil	60	1
2004-05	60	2	50	1	60	2
2005-06	60	Nil	41	Nil	60	1
2006-07	60	1	50	Nil	60	4
2007-08	60	3	50	Nil	58	3

Source: Office of the College.

With reference to the choice of faculty, it can be seen that in a small place like Dapoli, about 500 km away from Mumbai, greater number of Muslim girls have joined Commerce than Arts. Similar was the experience at a College in Chembur, Mumbai where more Muslim girls are studying in Commerce than in Arts. Thus, choices of Muslim girls with reference to faculty are showing a positive shift and resembles a general contemporary trend whereby increasing number of girls are now joining the Commerce faculty while traditionally they were concentrated in Arts faculty.

Table 4.9: Faculty-wise Enrolment of Girls in Belose College of Arts and Commerce, Dapoli

Year	*Arts*		*Commerce*	
	Total No. of Girls	*Total No. of Muslim Girls*	*Total No. of Girls*	*Total No. of Muslim Girls*
2002-03	125	04	47	18
2003-04	139	06	57	11
2004-05	154	09	51	10
2005-06	174	13	63	10
2006-07	184	11	80	20
2007-08	139	08	89	20

Source: Office of the College.

The particular trend as is observed in terms of choice of courses needs further detailed study. However, rather than affordability it could be the taboo on women's labour force participation which is still very strongly prevalent among this community that may be responsible for this. In the entire group, only 1 respondent reported the status of mother as working. In a couple of other cases elder sister was working. In two of the cases, the father had expired and the families were being looked after by uncles and elder brothers. The work participation of women was reportedly very low. Most of these respondents reportedly would be the first earning women in their families if allowed to work by the families.

The girls who are currently enrolled in these professional courses themselves are not very sure whether they would be allowed to take up a job later on. Like Fatima, doing B.Sc. IT, is already engaged to a boy from Lucknow whom she has not even seen or spoken to. The boy has studied up to 12th. Fatima does not know whether she would be allowed to work. Then why has she chosen a professional course like B.Sc. IT? To this, her answer was, "I am good in Maths and my teacher told me that I will have better prospects if I do B.Sc. IT".

Similarly, the girls who have taken admission to these courses in these colleges, come from diverse economic background from lower middle to upper middle class, most of them come from the families of self-employed income earners. This means that most of them come from families with some kind of business. The businesses ranged from caterers, decorators, textile embroidery, textile exports, electronic small shops, etc. But on the whole, most of them came from well-to-do families. However, it is usually observed in India that the women from business families and business communities are relatively less likely to take up or are allowed a salaried employment. This is true for all communities, though this situation is also changing now.

Thus, on the whole it can be said that while most of the respondents in the entire group came from families with relatively lower educational backgrounds, the respondents from professional courses had relatively better family educational backgrounds. There thus seems to be some correspondence between parents' education background and the choice of professional courses by Muslim girls.

The Muslim girls who have taken admissions in general B.A., B.Sc., B.Com. courses also came from diverse economic backgrounds. Many of these students have either their brother or a father working in some middle-east countries. Affordability, therefore, was not a major problem in many of these cases. Rather than affordability it is their social background which better explains the choice of courses. Those among the respondents who did express the possibility of taking up a job preference was more for teaching jobs. These factors are likely to explain the almost negligible presence of Muslim girls in undergraduate professional courses.

Perception of Benefits of Higher Education

In their book "India: Economic Development and Social Opportunity", Sen and Dreze (1999) point out that while for men, education motivation of the family/parent is not a problem, except for very poor families where loss of wage labour can be a direct disadvantage, for female education, this is often not the case. It is the combination of parental apathy, socio-cultural norms, direct and indirect costs, market failure to capture the true cost of girls' education, the law and order situation, inadequate and inappropriate educational infrastructure as well as low women participation in educational system that has made female education trail behind that of males.

In the face of such trends for girls' education in general and a few typical additions of Muslim traditions, what was the consideration of these girls in the fold of higher education? Almost all girls were unanimous that higher education gives them confidence and a sense of independence. Maybe, they will not be allowed to work in ordinary course, but thought that it is definitely going to be useful in any adversity.

But then what are the advantages of higher education if not allowed to work? They were specially asked a question as to in what way they are better off than their mothers. As mentioned earlier, the level of education of the mothers on the whole, was very low. In the respondents' view their higher education has helped them have better understanding of their mothers' conditions and thereby overcome the stress and strain they had experienced in their relationship with their mothers. Many of the respondents felt that their higher education will help them in familial roles as daughters and future wives and mothers. So even if they do not get the opportunity to work, the personal and social benefits of higher education would certainly make it worth.

Most of them expected to get married after their graduation though many expressed the desire to continue to postgraduation. About 20 per cent were determined to continue to postgraduation level. Some expressed that they will have to persuade their parents and overcome their resistance, but were hopeful to succeed like they have succeeded so far.

Roshia, for instance, who did her graduation in Economics in June 2007 and stood 1st in her College with 64.73 per cent marks in the university exam in April 2007, is quite uncertain and yet very positive and hopeful. She has got the permission from her father to continue to do her postgraduation or any other course for two more years, but not beyond that and certainly no permission to do a job. The father owns a small fruit and juice shop in the business centre in a commercial complex in Chembur; her elder sisters are married. They are educated up to 10th/12th.

Roshia is the youngest among 4 sisters. With the dream in her eyes, she says,"I probably may not get a chance to work or pursue the career. But right now since I have got the permission to study for two more years, I want to know what I can do further."

While narrating the advantages of higher education, many of them said that in order to compete in society and move ahead higher in life, education is a must. This suggested that not only motivational level but aspiration level also was unmistakably present, which could be treated as a positive and healthy indicator of emerging social pattern. It was, therefore, felt that the demand side factors like motivation, aspiration, security needs and demonstration effect and economic affordability are acquiring momentum and pushing the enrolment of Muslim girls for Higher education. The supply side lacuna, like availability of conducive social educational infrastructure may still exist. A large number of these girls came from Urdu medium schools; so the availability of good Urdu medium schools and junior colleges may further help in strengthening the positive forces generated by the demand side factors.

Socialization and Public Space Management

It has been argued that public space is gendered (Ranade, 2007). Comparing men and women of the same class and community, men have better access to public space at all times of the day. Women have to manufacture an appearance of purpose to legitimately access public space. Therefore, in their everyday occupation of public space, women are almost never found standing or 'loitering' in a public place without a clear purpose. The purpose defined movement of women is deeply connected to notions of respectability. Age is an important characteristic in determining access to public space. So in their late teens and early 20s women often get access to public space via medium of education.

The responses of Muslim girls did bring about this fact very strongly. While Mumbai is considered a very safe city, almost all girls mentioned that there are restrictions on their going out in the evening and almost no permission for coming home late after 6.30-7.00 p.m. Many respondents from Dapoli said that they are not allowed to go to the classes and those who are allowed, have to choose the classes that are close by and the timings are either morning or afternoon but never late evening.

The girls who stay in Muslim dominated localities said that while their parents may not mind little more relaxation and freedom being given to their daughters in the matter of socializing, going to public spaces etc., 'others' are watching and they pass comments and, therefore, parents are more rigid and careful. But the girls who are staying in mixed localities said that they do not feel such pressure of 'watchful' eye and also enjoy little more freedom. The 'mohalla' factor is present to a great extent in the restrictive environment experienced by many not only in terms of mobility but even in general.

Another important observation that emerged in this context was about the role of brothers in the family. Some of the girls said that they have not seen the college campus fully nor have gone to the canteen because of the fear of being seen by brother, or his friends and then reporting at home. According to some girls, rather than parents, brothers are stricter on them due to the fear of loosing the reputation in the eyes of their friends or risking the loss of honour if the sisters' names are 'discussed' among his friends etc. The 'brother' is thus treated as a 'surrogate patriarch' and, therefore, is obeyed dutifully by many girls even if the brother might be younger to them. While all girls did not agree to this, many girls reported this as their experience.

Islam does not permit a woman accepting maintenance from her ex-husband after divorce according to most girls. Similar is the view held by many experts. In fact, during group discussion at one of the colleges run by Muslim educational trust a opinion by the participants on this matter was very striking. The group consisted of students studying under professional courses like BMM, BMS and B.Sc, IT. When asked about having a law passed in the Parliament allowing all divorced women irrespective of class, caste, religion to claim maintenance from ex-husband, girls unequivocally and unanimously said that they will not support it and even if the law is made, they will not accept any maintenance if it is against the preaching of Islam. But many girls both in Mumbai as well as in Dapoli strongly felt that their education will help them stand on their own feet and give them independence. They thus look at education as a tool that lends confidence, self-dependence and, therefore, sense of empowerment.

Many said that other than coming to college, they hardly go out. This was the response of the girls who came from relatively poor households. Otherwise, many said that they do go out for shopping etc. but never alone. They generally never go out just for a walk. Usually it is either some social visit or shopping which means that other than college they go out mostly with family.

While the responses of the students interviewed strongly suggest continuation of a lot of restrictions, even a casual observer of Mumbai cannot miss the presence of them in all suburbs at all times and in many public places like restaurants with friends and families which was not the case in the past. Even single Muslim girls is confidently moving about which, of course, is a very positive and healthy change.

On Marriage and Social Interaction

While it is true that the girls' consent is a must at the time of *Nikaah* (marriage), they are actually given less freedom to choose their partner. Most girls expressed the desire for self-dependence or financial independence in future. Talking about their future, the discussion would shift to the choice of marriage partner and their wishes being respected in this regard. Out of the 101 respondents, 8 were already engaged. The family chose the boy; the girls had not talked to the boy, in two cases the girls had not even seen the boy.

One of them, Fatima, was doing her B.Sc. IT at a city college and was engaged to some boy from Lucknow. She had neither seen the boy nor spoken to him. The boy's family had some business and was economically alright. Fatima, eldest among 7 siblings said "well this is how things happen in many communities, at least in my family". When asked, "You are pursuing a professional course like IT, do you expect to be allowed to work after marriage?" And, her answer was "I do not know". "Have you been to Lucknow any time before?" "No." Well this may be a very extreme case, but most girls reported that selecting a life partner is not the freedom granted to them, while many said that they are asked and consulted, but it is a mere formality at the end, the parents' wish prevails.

One observation that emerged during discussion is worth noting. Some of the Muslim education trust colleges may be making special efforts to broaden the horizon of the Muslim girls. As one of the professors mentioned during the discussions, they really have to keep telling the girls that there are many things beyond marriage. "What is there, everybody gets married, but try to do something more than that in life" is what very often they tell their girl students. In fact, the professor observed that many of the girls get engaged by the end of first or second year. When the news of a friend's engagement comes the other girls feel sorry and worried. "So very often we have to underplay the importance of marriage and try to give them a broader vision and show new horizons", said the professor.

Incidentally, in the course of the group discussion, it was felt that many of the typical problems commonly affecting Muslim girls' lives can be better addressed in such colleges, since both the students as well as large number of staff form a much more cohesive group. In both the Muslim trust colleges, this undercurrent was felt. Of course, there is also another dimension to it that some outdated traditional practices, values and role models may get reinforced. As against this, in the non-Muslim trust colleges that were visited, while no Muslim specific issues were directly taken up, multi-cultural exposure may set in cross-cultural impacts, which will also have the positive effects on all concerned, since 'give and take' principle may naturally emerge. At conscious level there was less 'we and they' divide in such colleges. This was also the specific experience expressed by the Muslim girls from non-Muslim run colleges talking about their friends.

Polygamy was reported by three respondents, where the 'second mother' and her children were all staying together. Unexpectedly, and interestingly two of the respondents said that they are more attached to the second mother than their own mother as the second mother was more educated (up to 12th) more 'smart' and capable. One of the respondents said, her second mother is her role model because of her confident dealings with the outside world and even within the family. But many said that Polygamy is more common in Middle East countries than among the Muslims in India.

Invariably then the discussion would turn to Muslim Personal Law, particularly issues related to *talaq* and maintenance. In a group discussion on this issue, it so turned out that nobody had heard about *Shah Bano's case* and the Muslim Women (Protection) Act, 1986. But they all said that Islam takes their care because women are entitled to get 'Meher'— the sum of money agreed upon at the time of marriage and they also get *'Iddat'* (Maintenance for separation period of ninety days) and after that "Why should woman take anything from her ex-husband since now there is no relation between them?"

Another important observation was that the respondents from Muslim Trust College were more against any modifications or changes in personal law, while girls from non-Muslim trust colleges showed a great degree of flexibility. Whether it was the result of their greater exposure to cross-cultural interactions or whether they are basically more flexible and free so that they have joined these colleges could not be ascertained. But the girls in Muslim run colleges had come from Muslim dominated localities, while many girls who were studying in other colleges had come from mixed localities, though both these areas show a strong cluster or predominance of Muslim community like Kurla east and Cheetah camp at Mankhurd.

What is still striking was the response of the respondents from Dapoli, a small Taluka place away from the 'modern' and 'progressive' cosmopolitan city of Mumbai. At the group discussion, the girls said that if there is any such law passed, they have no objection to it,' as it would help the women who have no other means of supporting themselves, and have to fall back upon their parents or brothers etc.

Legal Literacy

The legal literacy among the girls was found to be extremely low. None of them was actually very sure, about their economic rights like right to property, right to share in the parent's property, or in the husbands' property etc. The response usually was "I think we have." Only a couple of them had heard about the *Shah Bano's case*, but did not know what it was all about. In all such matters, they simply seem to have complete faith in what has been told to them, as accepting the laws and provisions of Quran was unquestioning.

In this context, and particularly with reference to Muslim girls, the much debated issue, about the system of triple *talaq* invariably came up in the discussion. All the girls were aware of the system. They explained what happens in the interim period before final *talaq* happens. The entire process according to some is punctuated at the intervals of three months while some said at the interval of thirty days. So often within ninety days *talaq* is done according to them. The interim intervening period is given for, the girl to realize her mistake and to improve her behaviour; if she is not able to do this, then *talaq* happens. The tone of the narration of almost all of them, as quoted above, was that the girl has to improve and it was accepted quite in stride without questioning as to whether the husband also needs to correct himself! Of course, everybody believed that in a situation like this, where the husband is at fault, the girl can ask for *'khula'*. While they were aware about the provisions in Quran with reference to women's rights, they were not much well informed about whether in reality they get all the rights effectively and easily. On pointing out discrepancy between scriptures and practice they simply responded saying that probably implementation machinery needs to be improved upon but they all firmly believed that Quran has given all and equal rights to girls.

Burqa and Peer-Group Interaction

When asked about the inter-community peer interactions the respondents reported that most of their friends wanted to know why they wear *burqa*. Incidentally one observation is that prior to 1990s, the enrolment of Muslim girls in the colleges was relatively less as compared to the present, but a very small percentage among them actually wore *burqa* and came to college. While at present, the enrolment has gone up but so is the percentage of students wearing *burqa*. This may be indicative of changing composition of economic and social subsets within Muslim communities who are increasingly coming forward to send their daughters to colleges.

Does *burqa* come in the way of their peer interaction and socialization in the college? The reply is a 'No'. *Burqa* apparently did not pose any obstacle at the peer level. The girls belonging to non-Muslim communities also did not report any reservation in making friends or interacting with girls wearing *burqa*. The response thus, indicated that it is treated as private personal choice of dress code and did not really affect peer interaction.

But is it really a matter of private personal choice for the girls who actually wear it? There was a mixed response to this question.

Incidentally, this is one area where a large variation in views and practice was observed. Some respondents said it was compulsory and given a choice they would not like to wear it, while the others accepted that it was their own choice, but as they knew that they would have to wear it so why to resist?" Some said that since everybody in the family wears it so they also took it in stride. Two of the respondents said they like it as it has become their "fashion statement". One of the respondents even reported that when she first wore it, she felt very blissful and as if she has now come in the "arms of Almighty". Some of the respondents who are currently not wearing the *burqa* said they will have to wear after their marriage.

There also was a subtle difference between the responses of students from cosmopolitan city colleges and coming from affluent households and other Muslim girls in suburban colleges and from lower income groups. In the former group at the time of group discussions, out of 15 respondents only two were wearing *burqa* though except for two respondents all others said that they wear it when they go out of the college. All these respondents stay in a hostel and at that time were in jeans, skirts or *salwar kameez*. The effort seemed more like insisting and giving a message that though away from home, and in spite of their dress, they religiously follow all the community traditions including wearing *burqa*. While for the other group, the feeling was one of docile dutiful obedience and surrender. The questionnaire given to them contained one question on their wish list for themselves. To this the respondents in the former group mainly wrote on freedom in selecting marriage partner, education like MBA and good jobs; while the respondents in the second group listed freedom from *burqa* and greater freedom in selecting marriage partner. What is more significant to note is that, whatever may be their reasons to wear *burqa*, all of them said that they feel more safe and secured while moving out on the roads and in the society because of their *burqa*. Many of them said that *burqa* saves them from unsolicited, cross advances of men that women suffer and have to bear with in the buses or trains or market places. Such advances in public places are not faced by them while in *burqa*.

The deteriorating standards of morality in the society are thus once again likely to reinforce the traditional restrictive code of behaviour and conduct as far as women are concerned. All those who genuinely believe in equality, liberty and freedom should get more active and vocal so that the hands of the clock are not turned back. At the same time, the law enforcement authority and civil society should take these matters with much more concern or effectiveness. Of course, it is the credit to and the spirit of women that against all odds, they are learning to stand up, and face and keep moving ahead but it is certainly the responsibility of the entire society to make this march easier for all the women, not alone the Muslim women.

All the respondents except Bohra Muslim girls reported a regular and considerably strict adherence to all the practices expected of them like the daily routine of five times prayers, keeping fast during Ramzan etc. It was further reported that the ladies are not allowed to go to Masjid (most of Sunni Muslims). Many of them said that practices vary at the intra-community level due to the different interpretation of Quranic law, by different authors or Hajrats (Imams).

One of them, Shabnam, the elder of the two sisters, was very spirited. While sharing her experience with other friends from other communities or other people, she indicated her annoyance and displeasure about remarks made or comments passed about it even if they may be well-intentioned. She stays in Ghatkopar, a central line suburb of Mumbai. In 2003, there was a bomb blast in the public transport bus at Ghatkopar depot. At that time, it was suspected that one *burqa*-clad person had planted the bomb in the bus. After the incidence, one of the teachers advised her not to wear *burqa* so as to avoid suspicion. Similarly, somebody joked about it which caused a lot of displeasure and annoyance to her. Otherwise usually *burqa* does not come in the way, since all the friends at college are used to it and accept it in stride.

Incidentally, some very interesting experiences were narrated by some other college students. According to one respondent, one of her friends is working in one of the largest and high profile malls in the city. She wears the formal dress like trousers and shirt. She is working as a counter attendant at the mall. While on the job she has to interact with variety of customers. She wears *burqa* but removes it during duty hours and wears it again while going back home. According to her, the parents had put the condition of wearing *burqa* for giving her permission to work. Further, according to this girl, there is hardly any chance that any of her neighbours or known people will go to that mall since the mall is far away from the place of her residence. She, therefore, feels safe to remove her *burqa* while on the job! In their opinion, people from other community are unnecessarily 'high stung' on *burqa*. Most girls who wear it are fine with it. They not only do not mind wearing it, but actually say that they wear it out of their own choice.

In this context one incidence narrated by one of the lady professors is very interesting. Once in her college one of the Muslim girls, who wears *burqa* while coming to college, used to remove and kept in the desk during class hours, lost her *burqa*. Somebody had probably stolen it. This was quite a panic situation for this girl not knowing how to manage going home without *burqa*. Her friends then collected some money, went out and bought a new *burqa*, after which only this girl could go home.

Why would any young girl want to cover herself totally and hide her identity? Is it not more natural that at young age one would want to wear nice dresses, jewellery, accessories and show it around to friends? "Well, we of course wear nice dresses and jewellery as well as other accessories. But as per Islam, decorating ourselves (*sajna-dhajna*) is only for ourselves and for our husbands and not for anybody else. And we strictly believe in Islam."

Sana follows *purda* very strictly. Throughout our two hours discussions, only a line showing pair of eyes was visible, otherwise the girl was fully clad in *burqa;* even her hands were covered with full sleeves and she was also wearing hand gloves. At the time of discussions with the group, one of their lady professors Mrs. Khan was also present. The professor wears full sleeves *kurta* and *salwar* and covers herself with *dupatta* but does not wear *burqa*. She certified that the way Sana does *purda* is the perfect; nothing except eyes should be seen.

But does it restrict their mobility, participation in extra-curricular activities etc.? According to professor Khan, Ms. Shirin who also wears *burqa* is one of their best elocutors. She always participates in debates, elocution etc. and have won prizes. Thus, this does not offer any limitation, for the one who wants to perform. The professor herself does not wear *burqa* or her daughter, studying in the same college at S.Y.B.M.S., who wants to become an IAS officer.

A professor from the college at Dapoli observed, "Here we cannot differentiate between Hindu and Muslim girls on the basis of *burqa*, because not many Muslim girls wear it. In fact, the differentiating symbol is a red or black *bindi* that the Hindu girls put on their forehead, but not Muslims! In S.Y.B.Com. class, there were 7 Muslim girls present, but no one of them wears *burqa*. However, it was observed that in their F.Y. class out of 17 students, 11 were wearing *burqa*! That was quite perplexing. When enquired no convincing explanation emerged. Their reasons for wearing *burqa* were: (a) because friends wear, (b) everybody at home wears, (c) I like to wear. There were two girls who said their mothers do not wear *burqa,* but they do because they like to. There is also a good amount of romancing about *burqa* too, as one of the girls said "It's nice and fun!"

The diverse socio-economic background of the migrant Muslims of Mumbai reflects itself in the diverse outcomes and aspects with reference to the degree of freedom enjoyed or rather 'allowed' to the girls. This is evident from varied examples in this sample of respondents. There are cases of Naznin and Alia, studying in high profile City College who wear short skirts and jeans, sport short boy-cut hair style. While Farah also studying in the same college, wears *Burqa* and had to struggle a lot to get permission to continue higher education. In this sample Atiya was some one with very well maintained manicured hands and feet, and Mehenaz who says that you can't pluck and shape your eyebrows because Islam does not allow women to cut their hair! Mumbai thus represents all these shades of thoughts and behaviour.

Mumbai Factor

Mumbai Violence and its Impact

The massive increase in violence in the last three decades in India brought home the fact that modernisation in general and economic development and the speed of the culture of modern science in particular, were no barriers to ethnic and religious violence. In fact, they sometimes stoked such violence. Most of the cities where "communal" violence occurred also were relatively more industrialized cities in that region (Nandi, 1999). The city of Mumbai has experienced a good amount of outbursts of violence in the past decade.

Some studies carried out by NGOs and independent scholars in post-Godhra Gujarat reveal that there is a substantial increase in number of Muslim students not only in higher education but even in medicine and engineering. The Muslim community seems to have realized a need to get better education to improve its economic conditions among other benefits. A similar trend is probably at work in Mumbai, post-rioting decade of 1990s.

"Mumbai changed after 1993" is a common phrase of long-time residents of the city. A decade later, subtle changes are evident in more than one way in Mumbai, in the attitude of Muslims towards the government and the political parties, and in the way *Mumbaiites* perceive their city.

Historically, Muslims of Mumbai have been one of the most heterogeneous groups in India. The oldest Muslim communities in Mumbai are the Bohras and the Khojas, who are relatively small, but wealthy trading communities of the Shia persuasion, and the Sunni Memons. The city also has a large number of Konkani Muslims, who hail from the coastal region south of Mumbai and constitute a Marathi group that is well-integrated into Mumbai's industrial economy. The majority of the Muslims of Mumbai have come from Uttar Pradesh and Bihar, in search of jobs. The Ansaris, who are Muslim weavers from Uttar Pradesh came in large numbers to work in the textile mills. Since the 1970s, there has been a steady inflow of impoverished Muslims from Bihar; they do many of the lowest-paid jobs, as coolies, sweepers and so on. Finally, there are smaller groups of Muslims from south India. The divisions between the older Muslim population and the newer migrants from north India are manifested not only through sectarian differences (Shia and Sunni) but also through class differences. The more recent migrants dominate the mill districts such as Nagpada and Madanpura, which have numerous small and informal businesses and workshops.

Eminent social scientist Asghar Ali Engineer emphasizes that politically, in the post-riot period, Mumbai's Muslims have been a subdued lot. He (in a personal interaction) says:

> "The present-day Muslim leadership is low key and more cooperative. There is a greater emphasis on the uplift of the community and women's education. Post-Gujarat, there is a sense of helplessness. This is different from the post-Babri period, where there was no sense of helplessness, only a deep sense of insecurity and a

> realization that confrontationist postures will not pay and that we have to correct our faults."

Journalist Jyoti Punwani explains: "The city's Muslims have refused to get provoked into coming out on the streets. They want no more confrontations with the police." At the same time, there is a preoccupation with making up for the lost years by excelling in academics—for the last three years, school and college toppers have been Muslims, many of them from lower-middle class backgrounds.

Mumbai's communal tension cannot be entirely explained either by the argument that Muslims and Hindus cannot co-exist or by the suggestion that the ghettoisation of the minority community is bound to create conflict. According to Ashish Nandi that communal riots in India are more on the lines of the proletariat-bourgeois riots during the industrial revolution in Britain. All these riots (communal riots in India) have happened in the industrial cities mainly and where Muslims are a relatively larger group. They are therefore more in nature of industrial riots between haves and have-nots and a communal overtone is one additional dimension to it.

The masses who participate in a riot are not a homogeneous lot. There are differences in their aims and motives. While some are members of political networks, which stand to gain through violence, others are lumpen elements who participate on their own. It is only proper awareness among people and active role of civil society actors which can help contain major mishaps. We need aware and vibrant civil society to contain outbreak of major communal violence. When civil society gets polarised on communal lines, it becomes very difficult for civil society to intervene.

Safe Place–Mumbai

In the opinion of most of these girls, Mumbai is the safest place. They like to be in Mumbai and would not want to settle anywhere else. Well it is quite natural to feel attached to the place one is born and brought up, but are there any special reasons? Do they have relatives in other parts of the country and have they experienced a difference in the living conditions? In all group discussions this was one of the areas where respondents not only overwhelmingly responded but also felt that they were really lucky and fortunate to be staying here.

"The difference?"

"Well, here nobody bothers about anybody, which is in a way good so that you can have more freedom."

"Like?"

"I can go out with friends of course I have to return before evening, but I can go once in a way....Life here is much easier with all amenities."

"But so it can be in many major cities or other metros."

"No ma'am Mumbai is Mumbai. I will tell you my experience, ma'am", said Shagufta.

Shagufta's family is originally from Hyderabad. "Last time when I visited Hyderabad, I was going out with my cousin sisters, we followed a particular road. While coming back home, I suggested to them that we will go by another road and immediately they said, "Are you mad? That is not our way that is their way and this is 'our' way! We have to stick to our way; nobody crosses each other's way". Shagufta was surprised at this and said, "In Mumbai we do not have any such thing. Anybody can go anywhere!"

Shagufta is studying in Third Year (TY) B.Com. Her father is working in Saudi Arabia. The mother with 3 daughters stays alone in Mumbai. The father visits during holidays etc. But usually it is the mother who takes all decisions since the father is away. The mother and Shagufta wear *burqa*, but younger sisters have not yet started. They are living in a cosmopolitan locality and in their colony theirs is the only Muslim family. So how is the nature of interaction with members of other communities? "Well, very friendly, absolutely no constraints or barriers. We go to each other's houses, share sweets on each others festivals and live typically like Mumbai neighbours! They are there in case you need them, otherwise you are on your own and nobody bothers much." In her opinion, this is the *'khasiyat'* (special feature) of the city of Mumbai. It is safe, truly cosmopolitan and pluralistic in nature.

What probably she experienced at Hyderabad was a kind of ghettoisation which is absent in Mumbai in spite of a very bad history of riots in the recent past. Many girls were not very articulate in their answers but felt that they 'feel' free in Mumbai, in spite of many of them having to wear *burqa*, having to put up with many restrictions.

There is very little "ghettoisation" as such in Mumbai, though there is clusterization (which is present not only on the lines of religion but also language and region) since many Muslim households are actually living in mixed cosmopolitan societies. Most of the girls except for the college located in Byculla came from scattered localities. However, it is still true that there is a good amount of clusterization, in the sense that there are geographical pockets where there is heavy concentration of Muslim communities; but their residential houses though in the same locality are mixed community colonies. The choice of the housing is more a matter of the class position of a household in a given locality. Another factor for clusterisation may be the migrant nature of this community. A large number of Muslims in Mumbai are migrated from Uttar Pradesh, Bihar, Andhra Pradesh, Tamil Nadu and Madhya Pradesh. They, therefore, stay in the areas where there is a larger concentration of migrants from their state. However, it is believed that after the rioting decade of 1990s, Mumbai's religious landscape has changed, with many households relocating themselves near to their own community households, and clusters turning into ghettoes (see Shaban, 2010).

Maybe to appreciate this, we need to move beyond mere symbolism and gesture. Freedom is in its air! All respondents from Mumbai felt that they 'feel' 'free' in Mumbai as compared to their native places. Even those who actually are subjected to a lot of community pressure, patriarchy and gender bias, strongly feel that they are much better off and feel more privileged than their cousins and other girls in their native places and in most cases it is due to the fact that they are allowed to attend colleges.

While it is true that divergent views exist, it is equally true that for a long-time resident of Mumbai, there is a "sudden explosion" in the number of *burqa* clad women and girls all-over the city and particularly on the college and university campuses. Further, for a new entrant like one Professor who came to Mumbai from Baroda after her marriage 4 years before, it is surprising that one sees so many *burqa* wearing females all-around which is not the case back home in Baroda. Not that there is less number of Muslims there, according to this Professor, but probably the practice of wearing *burqa* is less.

A similar observation was made by a professor at a College of Commerce, Dapoli. According to him, Muslims in Konkan region are economically better off and socially quite an upwardly mobile lot. Many of them have their male members working in Middle East. As income is not a constraint, they send their children (including girls) for higher education.

On the whole, there is one more motivational factor for the Muslim girls to take up higher education and that is they are likely to enhance their status to avail a match working abroad. As stated earlier, many families prefer their daughters attend colleges, which are close to their residence. Mumbai is a place where it facilitates and satisfies this condition to a large extent. In the last 10 years, the number of colleges under Mumbai University has increased substantially. There are 400 colleges in Mumbai region. The availability of colleges relatively nearer to their homes may make it easy and convenient for them to send their daughters for higher education. As mentioned earlier, the local government runs over 200 Urdu medium schools, which provides initial facilitation for the education. In this context, it may also be noted that there are about 10 general stream colleges being run by Muslim educational trusts. Thus, on the whole, the educational infrastructure in Mumbai is one of the major facilitators.

Urdu: Mother Tongue by Default?

Unexpected information that emerged from this survey is about the status and role of urban language. In fact, initially while conceptualizing this study 'Urdu' or for that matter language as such was not factored in. Information on the mother tongue was sought only as a matter of their 'bio-data' profile. Further, at the most it was expected to give some information on the region of India they hail from. However, during the survey some very different and unexpected perspective emerged particularly about Urdu language.

Mumbai has 11,99,441 Urdu speaking persons as per the census report, and 15,33,166 have reported their religion as Islam (Census of India, 2001). Most of the Muslims in Mumbai are migrants; inter-state or/and intra-state. Surprisingly, however, 101 of the respondents interacted with only 15 had given their mother tongue other than Urdu. Even if the respondents' native place is Hyderabad, Chennai, Ratnagiri or Kolkata, they said their mother tongue was Urdu. Actually, it is usually expected that the region from where your family hails is likely to be your mother tongue. In discussion with one of the groups where this matter was being discussed, a respondent said that her family is from Karnataka. Her father first came to Mumbai from Bangalore. Her mother's family came to Mumbai from Pakistan much before partition. So after coming to Mumbai the families followed Urdu as their main language. Probably, what might be the case that the Muslims having come to Mumbai from different places have found knowledge of Urdu as a common factor establishing social relationships with other Muslims and therefore over a period of time Urdu has been adopted by the families.

In this context, it is also worth noting that almost all respondents reported that their families do arrange some sort of training or orientation in Islamic preaching and scriptures through *Ulema* or some *Moulavi*. The knowledge of Urdu thus becomes necessary. Or at least it helps. The Brihan-Mumbai Municipal Corporation (BMC) run Urdu medium schools enabled them to put their children in them which further increased the utility of the language and so in the course of time, the language. The language has also given them a strong sense of identity and bonding. One respondent who came from Malegaon, Maharashtra, mentioned that many of her male cousins have done D.Ed. as it becomes easier to get jobs in Urdu medium schools. This respondent's family has shifted to Mumbai and she is the first member in the family to attend the college.

While the language might have generated some bonding among the Muslim community from different parts of the country, usually for marriage they prefer their regional Community, i.e. to say a Bihari Muslim will marry into a Bihari Muslim family. However, many respondents said that these days the families only insist on the religion but not regional community. So will they be allowed to have a love marriage? Most said "No". But those who responded positively said that "Provided the boy is also Muslim". Inter-religion marriages are thus not acceptable. Some girls said that a boy can marry a Hindu girl because after marriage she automatically will follow her husband's family name and religion. On being pointed out that actually a girl may choose/retain her original religion if she so desires, they said they were not aware about it and anyway it mostly does not happen that way.

Conclusion

There is an increasing acceptance of women's higher education among the Muslim community. The parents across different levels of the income, occupation and education are on the whole supportive. There are a large number of first generation learners who are coming from a diverse occupational background from small skilled and unskilled workers, petty traders and small to medium size businesses into the fold of higher education. However, the traditional family gender roles continue to dominate the expectations of parents with reference to their daughters though they have conceded to their demands of higher education. In certain cases and in the eventuality of parental resistance, the other supports like teachers and NGOs have been a major and positive factor. The self-motivation level among the young Muslim girls is very high. While the course choices are still traditional, there are noticeable shifts therein. Availability of large number of Urdu medium schools (private and public) in Mumbai seems to have helped at the initial levels of education. The aspirational level also is unmistakably present. Higher education is seen as insurance against uncertain future by some. Migration remittances of father and other family members have a positive impact on women's education. *Burqa* is being used by both, the families and girls, to negotiate and seek greater mobility and freedom. Legal literacy is low, across all communities particularly women rights literacy.

Mumbai impact has been mainly in terms of widespread better educational infrastructure, cross-cultural demonstration effect in terms of women's education and employment, and a sense of safety and security.

Thus, the educational canvass of Muslim girls in Mumbai is changing. The signs of modernization and development are unmistakable. They are changing everywhere, not only among the Muslim community but among other marginalized and non-marginalized sections also. It is our collective responsibility and a common destiny that we as a society must ensure that the signals of positive changes are picked up, analysed and consolidated to shape better future. The Muslim girls have definitely taken their early baby steps, which hold a good promise for the journey to continue.

REFERENCES

Ansari, Ashfaq Husain. 2007. *Basic Problems of OBC & Dalit Muslims*, New Delhi: Serial Publications.

Census of India. 1991. *Population Census 2001*. Data on CD. Office of the Registrar General and Census Commissioner, Ministry of Home Affairs, Government of India, New Delhi.

Census of India. 2001. *Population Census 2001*. Data on CD. Office of the Registrar General and Census Commissioner, Ministry of Home Affairs, Government of India, New Delhi.

Engineer, Asghar Ali. 2001. *Islam, Women, and Gender Justice*, New Delhi: Gyan Publisher.

Government of India. 2006. *Social, Economic and Educational Status of the Muslim Community of India* (Sachar Committee Report), Prime Minister's High Level Committee, New Delhi.

Hasan, Zoya and Ritu Menon (ed.). 2005. *In a Minority—Essays on Muslim Women in India*, New Delhi: Oxford University Press.

Hasan, Zoya and Ritu Menon. 2004. *Unequal Citizens: Study of Muslim Women in India*, New Delhi: Oxford University Press.

Hasan, Zoya and Ritu Menon. 2004. *Educative Muslim Girls: A Comparison of 5 cities in India*, Delhi: Women Unlimited.

Mayaram, Shail, M.S.S. Pandian and Ajay Skaria, (ed.), 2005, *Subaltern Studies XII: Muslims, Dalits and the Fabrications of History*, New Delhi: Permanent Black.

Nandy, Ashis. 1999. "Coping with the Politics of Faith and Cultures: Between Secular State and Ecumenical Traditions in India", in Joanna Pfaff-czarnecka *et al.* (ed.) *Ethnic Futures*. New Delhi: Sage Publications.

Ranade, Shilpa. 2007. "Dangerous Liasions", *Economic and Political Weekly*, 42(7).

Sen, A. and John Dreze. 1999. *India: Economic Development and Social Opportunity*. New Delhi: Oxford University Press.

Shaban, A. 2010. *Mumbai: Political Economy of Crime and Space*. Hyderabad: Orient Blackswan.

Siddique, Aboobacker P. 2005. "Panchayati Raj and Women in Kerala: The Case of Muslims", in Zoya Hasan and Ritu Menon (ed.), *In a Minority – Essays on Muslim Women in India*. New Delhi: Oxford University Press.

Vatuk, Sylvia. 2005. "Muslim Women & Personal Law", in Zoya Hasan and Ritu Menon (eds.), *In a Minority – Essays on Muslim Women in India*. New Delhi: Oxford University Press.

5

Poverty, Unemployment and Housing Distress among Muslims in Mumbai

Abdul Shaban

Urban centres in the country are undergoing significant social and economic changes and these changes are producing and reproducing new geographies of marginality and centrality in cities/towns. The urban centres have also seen considerable seasonal and permanent migration of rural population (from rural to urban centres). In this migration process, the likelihood of migration of the communities having less productive assets in rural areas and artisan class remains high. It is not only that urban areas offer better employment opportunities but also better wages. However, migration may also be due to push factors like security issues. These pull and push factors, among others, determine the spatial flows of population. Muslims in rural areas in many parts of the country are faced with livelihood challenges because of lower productive assets ownership (like land) and also in recent years they have experienced increased marginalization due to sectarian politics and culture. Additionally, as a significant number of Muslims are engaged in artisan activities, the government run programmes for unskilled workers remains not only unattractive to them but discriminatory practices in implementation also make these programmes inaccessible to them. Cumulatively, these factors push larger number of Muslims to urban centres. In fact, in many of the states like Maharashtra Muslims are now largely concentrated in urban centres while from states like Uttar Pradesh and Bihar most of them engage in seasonal migration to urban Centres.

Historically, Mumbai has been one of the major urban centres in the country drawing a significant number of Muslims from rest of the country. Muslims have lived in Mumbai for more than 700 years (Khan, 2007), but as the economic significance of the city has grown in recent years, the migration of Muslims to the city from various regions of the country has also increased. As per the Census of India 2001, Muslims constitute 18.56 per cent of the total population (11.98 million) of Greater Mumbai. The Muslim community in the city is very heterogeneous. Like any other community in the city, it is divided on the bases of region, language, sect and caste. The religious identity of the community is constructed more externally than is emerging from the homogeneity of differing groups within, or the Muslim communities themselves like to claim. Ironically, these externally imposed supra-identities undermine their particularities. The homogeneity thus constructed and supra-identity manufactured, often determine their fate with respect to the public provisioning and decisions by the other communities (as in riots, all Muslims are treated one). The present essay examines income, poverty, asset ownership, access to banking and credit facilities, employment and housing distress among Muslims in Mumbai.

For this study, data have been collected through a household survey. In the absence of area-wise authentic statistical details on the Muslim population within Mumbai, seven areas were selected for the purpose. These areas were Govandi, Kurla, Bandra (E), Bandra (W), Dindoshi, Byculla-Nagpada (B-N) belt and Dongri. These areas were selected with a view to capture variances existing among the Muslims of Mumbai especially with regard to concentration of population, sects/communities, and socio-economic status. Govandi, Bandra (E) and Dindoshi are lower class areas while Dongri and B-N belt are middle and upper middle class localities. Bandra (W) represents dispersed middle and higher classes of Muslims. In India, Muslims are divided into two major sects: Sunni and Shia. However, within these two sects, there are a number of sub-sects/communities. In Mumbai, distinction among different communities like Bohra, Khoja, Memons and rest of the Shias and Sunnis are quite prominent at the organisational and interactional levels. Bohras and Khojas belong to Shia sect while Memons to Sunni sect. Within Sunnis there are many communities who belong to Other Backward Class (OBC) (the survey could not find any family belonging to OBC from Shia sect). Given the heterogeneities, the discussion in the chapter are organised along the comparative analysis of these sects/ communities and localities. The sect/community-wise figures presented in the chapter are exclusive of each other. Total 1,643 households (9362 persons) were covered in the survey, which was carried out during June to November 2007. The organisation of the rest of the essay is as follows. We, first, examine the economic status, especially income level, household asset and access to banking/credit facilities, followed by employment and unemployment of individuals and their distribution in various occupational categories. House ownership, household amenities, and living environment are examined next. The last section presents conclusion of the study.

Economic Status

Income and Poverty

A significant proportion of the total Muslim households in the country lives in poverty or at the verge of poverty. The National Sample Survey Organisation (NSSO) 61st round (2004-05) data based on 365 days reference period shows that, as per head count ratio, 31 per cent of the Muslims live below poverty line as compared to an overall poverty level of 22.7 per cent in the country. Muslims in urban areas have the highest level of poverty (38.4%) among all socio-religious groups (other socio-religious group being SCs/STs, Hindu OBCs, Hindu General) (Government of India, 2006). The present study shows that in Mumbai 29.1 per cent of the families (34.1% of the total surveyed population) live below the official poverty line income of Rs. 591 for urban areas. In comparison to this, a recent survey by the Government of Maharashtra indicates that only 8 per cent of the families in Konkan Municipal Councils live below poverty line (Rs. 591 or less). The overall household poverty in 156 Municipal Councils in the state has been reported to be 21 per cent (*Times of India*, 2008).

Within Mumbai, the magnitude and expanse of poverty among Muslims varies a great deal. In Govandi, 41.3 per cent of the Muslim families live below poverty line and the same in Kurla and

Table 5.1: Per capita Monthly Family Income by Locality (per cent)

Locality	*Per capita Family Income (Rs.)*							
	Below 591	*591 to 1000*	*1001 to 1400*	*1401 to 2100*	*2101 to 3500*	*3501 to 5000*	*above 5001*	*Total*
Govandi (N=375)	41.3	33.3	9.6	10.1	4.0	.5	1.1	100
Kurla (N=388)	35.6	35.8	9.5	8.8	5.9	1.8	2.6	100
Bandra (E) (N=169)	27.2	36.7	14.8	12.4	5.9	1.2	1.8	100
Bandra (W) (N=82)	22.0	32.9	8.5	4.9	2.4	2.4	26.8	100
Dindhosi (N=123)	27.6	42.3	16.3	9.8	4.1			100
Dongri (N=193)	16.1	42.0	10.4	15.0	9.3	5.2	2.1	100
B-N belt (N=313)	17.9	44.4	13.4	8.0	8.0	2.9	5.4	100
% of total no. of families	29.1 (N=478)	38.0 (N=625)	11.4 (N=187)	9.9 (N=163)	6.0 (N=98)	1.9 (N=32)	3.7 (N=60)	100 (N=1643)
% of total persons	34.1 (N=3190)	35.6 (N=3336)	11.1 (N=1036)	9.0 (N=838)	5.6 (N=521)	159 (N=1.7)	282 (N=3.0)	100 (9362)

Source: Based on Household Survey.

Dindhosi are 35.6 per cent and 27.6 per cent respectively (Table 5.1). B-N belt and Dongri (less than 20% each) have minimum level of family/household poverty in the city. It is also shocking to note that a significant proportion of the Muslim families in all the localities live on the verge of poverty: with per capita monthly income between Rs. 591 and Rs. 1,000. Overall, 38 per cent of the families in Mumbai live within this range of income. Thus, more than two-thirds of the sampled Muslim families in the city live in extremely adverse economic situation (below per capita income of Rs. 1,000). Also, the share of families having per capita income above Rs. 5,000 is extremely small, i.e. 3.7 per cent. Locality-wise only Bandra (W) has a sizeable proportion of the Muslim families with per capita income above Rs. 5,001 (about 27%). In Dindhosi, none of the surveyed families has per capita income above Rs. 3,500. In the context of measuring poverty by income criterion, it must be noted that Rs. 591 is too small an amount. In one of its recent study on 'Impact Assessment of Resettlement Implementations for Mumbai Urban Transport Project-II (MUTP-II)' for the World Bank and Mumbai Metropolitan Region Development Authority (MMRDA), Tata Institute of Social Sciences (Sharma *et al.*, 2008) has used monthly family income of Rs. 3,000 as poverty line income. As per Rs. 3,000 monthly family income, the number of Muslim families below poverty line rises to 41 per cent.

Among various sects/communities within Muslims, poverty is highest among Sunnis (both general and OBC). As against the city level family poverty of 29.1 per cent (Table 5.2), Sunni (general) and Sunni (OBC) have household poverty of 30.7 per cent and 30.3 per cent, respectively. The least level of household poverty is among Bohras (2.6%) followed by the Khojas (8.3 %). Among Shias and Memons, the household poverty is 17.4 per cent and 19 per cent, respectively. It is interesting to note that most of the Memon families of the city are living in Bandra (W), where the share of the families below poverty is the least. Besides Memons, Bohras are the richest community in the city among Muslims.

The average per capita monthly income of the families in the city is Rs. 1,481. As against this, the per capita family income in Dindhosi is only Rs. 922. Govandi has family per capita income of Rs. 976, followed by Kurla with Rs. 1,139. Bandra (W) has the highest per capita income, i.e. Rs. 6,663. The average per capita

Table 5.2: Per capita Monthly Family Income by Community/Sect (per cent)

Community/Sect	*Per capita family income (Rs.)*							
	Below 591	*591 to 1000*	*1001 to 1400*	*1401 to 2100*	*2101 to 3500*	*3501 to 5000*	*Above 5001*	*Total*
Sunni (gen) (N=1219)	30.7	37.6	11.6	9.4	5.4	1.6	3.7	100
Sunni (OBC) (N=294)	30.3	39.1	9.5	10.5	7.1	1.7	1.7	100
Memon (N=21)	19.0	47.6	4.8	9.5	4.8		14.3	100
Shia (N=46)	17.4	37.0	10.9	19.6	8.7	4.3	2.2	100
Khoja (N=24)	8.3	45.8	20.8	16.7			8.3	100
Bohra (N=39)	2.6	35.9	15.4	5.1	15.4	15.4	10.3	100
% of total family (N=1643)	29.1	38.0	11.4	9.9	6.0	1.9	3.7	100

Note: The figures for all the Sects/Communities are exclusive of each other.
Source: As for Table 5.1.

family expenditure varies among the localities. As against the overall average of Rs. 1,103, Dindhosi has the per capita family expenditure of only Rs. 678, followed by Govandi Rs. 690, Bandra (E) Rs. 861, B-N belt Rs. 1,117 and Bandra (W) Rs. 4,684. Table 5.3 provides expenditure group-wise classification of families across localities. It is seen that while about 26 per cent of the total families have per capita expenditure below Rs. 500, the share of families having per capita expenditure below Rs. 500 in Govandi is approximately 43 per cent. The same is about 15 per cent in Dongri and 14 per cent in B-N belt.

The per capita family income and expenditures also vary considerably across communities. While the average per capita income among the Memons is Rs. 8,342, it is only Rs. 1,132 among the Sunni (OBC), Rs. 1,413 among Sunni (General), Rs. 1,512 among Shias, Rs. 1,661 among Khojas and Rs. 2,411 among Bohras. The Memons have the highest per capita income, and spend (monthly per capita expenditures Rs. 7,236) more than other communities among Muslims. The Khoja and the Bohra communities each have more than 80 per cent of their families in expenditure class above Rs. 591, while about two-fifth of the Sunni families have per capita expenditure below Rs. 591.

Table 5.3: Per capita Monthly Family Expenditure by Locality (per cent)

Monthly Expenditure (Rs.)	*Locality*							*% of Total families*
	Govandi	*Kurla*	*Bandra (E)*	*Bandra (W)*	*Dindhosi*	*Dongri*	*B-N belt*	
Below 500	42.7	25.8	24.4	23.2	30.3	15.0	14.1	26.2
501 to 590	11.5	13.1	9.5	7.3	16.4	4.1	8.3	10.4
591 to 1000	28.3	36.1	39.9	25.6	37.7	33.7	38.7	34.5
1001 to 1500	10.9	13.4	16.1	12.2	13.1	25.9	21.4	16.0
1501 to 2000	3.7	4.6	5.4	3.7	2.5	8.8	8.6	5.5
2001 to 3000	1.9	4.1	4.2	2.4		8.3	5.4	4.0
3001 to 5000	1.1	2.3	0.6	4.9		1.6	2.2	1.7
5001 to 7500		0.5		4.9		2.1	1.0	0.8
7501 to 10000						0.5	0.3	0.1
10001 & Above				15.9				0.8
Total	100.0	100.0	100.0	100.0	100.0	100.0	100.0	100.0

Source: As for Table 5.1.

Access to Credit Facilities and Indebtedness

A large share (80.5%) of the Muslim families in the city has no debt; however, in this context, one finds variation among the localities. While, 28 per cent and 24 per cent of the families in Govandi and Dindhosi, respectively, are in debt, the share of indebted families in Dongri and B-N belt is only 13 per cent and 10 per cent respectively. Although the share of indebted families across the localities is low, the share of families having debt above Rs. 25,000 is high (59.5% of total indebted families) among the indebted families. Overall, the average debt in the sample Muslim families is Rs. 15,676: the highest being in Dongri (Rs. 26,438) and the lowest in Bandra (W) (Rs. 3,201). However, the average debt (computed exclusively for the indebted families) is Rs. 80,233: highest being in Dongri (Rs. 2,04,104) and B-N belt (Rs. 1,65,433) and the lowest in Bandra (W) (Rs. 37,500). The largest percentage of families having debt is from Sunni (General) followed by Sunni (OBC). Out of the total surveyed families of Sunni (General), about 20 per cent are in debt, while the share of indebted Sunni (OBC) families is about 19 per cent. Memons and Bohras have the least shares of indebted families; about 5 per cent and 8 per cent respectively.

The major sources of credit across the localities and communities are non-institutional sources (Tables 5.4 and 5.5). Overall 14.2 per cent of the families borrow from friends and relatives. Further, 1.9 per cent borrow from moneylenders. In comparison, the institutional credit (from banks and co-operatives) is extremely small across all localities and communities. Only 2.8 per cent of the total families have loans from banks and co-operatives. This may be interpreted as the fact that the Muslims find it difficult to borrow from formal institutions. The reasons attributed for the same are: (i) banks are less willing to finance Muslim families, especially in Muslim dominated areas called 'negative' areas (this practice becomes significant in view of the fact that the Muslim population in the city due to fear and other biases in housing market is ghettoised in a few locations), (ii) weak financial status makes Muslims unable to provide guarantors and collaterals required for loans by the banks, and (iii) low educational status and lack of awareness make it difficult for a majority of Muslims to apply for institutional loan. Data show that among the communities, higher proportion of Memons borrowed from the moneylenders. In fact, none of the surveyed families from Memon and Khoja communities claimed borrowing money from Banks or Co-operatives.

Table 5.4: Sources of Debt by Locality (per cent)

Sources of Debt	Locality							% of Total Families
	Govandi	Kurla	Bandra (E)	Bandra (W)	Dindhosi	Dongri	B-N belt	
Relatives/ Friends	21.3	15.7	16.0	3.7	18.7	10.4	6.4	14.2
Money-lenders	3.5	1.8	4.1	1.2		1.0	0.6	1.9
Banks/Co-operatives	2.4	2.1	4.1	3.7	6.5	2.1	2.2	2.8
Credit societies	1.3	0.8						0.5
No Debt	71.5	79.6	75.8	91.4	74.8	86.5	90.8	80.6
Total	100.0	100.0	100.0	100.0	100.0	100.0	100.0	100.0

Source: As for Table 5.1.

Table 5.5: Sources of Debt by Community/Sect (per cent)

Sources of Debt	*Community/Sect*						*% of Total Families*
	Sunni (General)	*Sunni (OBC)*	*Memon*	*Shia*	*Khoja*	*Bohra*	
Relatives/ friends	14.5	14.3	4.8	19.6	12.5	5.1	14.2
Money-lenders	2.1	1.4	4.8	2.2		2.6	1.9
Banks/Co-operatives	3.0	2.7		2.2		2.6	2.8
Credit societies	0.6	0.3					0.5
No Debt	79.8	81.3	90.4	76.0	87.5	89.7	80.6
Total	100.0	100.0	100.0	100.0	100.0	100.0	100.0

Source: As for Table 5.1.

About 65 per cent of the Muslim families have bank accounts. In light of the above discussion, this shows that Muslims face more difficulty in acquiring loans from banks than holding deposit accounts. In Dongri, B-N belt and Bandra (W), the share of families having bank account is above 70 per cent in comparison to around 48 per cent in Govandi. The lower share of total families having bank account in Govandi is due to the fact that banks require residence proof for opening a deposit account and Govandi being a slum area, people find it difficult to acquire the same. Further, this area is predominantly inhabited by poorer households where only some of the families are able to save. Community-wise, Sunni (OBC) have the least share of families (57.1%) and Bohras (84.6%) the largest share of families with bank accounts.

Given that most of the Muslim families are poor or have just survival income, lack of substantial savings is an obvious fact. In the long-run, the health and other exigencies eat away even the meagre saving which some of the families could have. It is found that about two-thirds of the families in the city do not save due to low level of income. In Govandi, about 48 per cent of the families are unable to save. However, among those who save, the preferred place of saving is bank (42.9% of the total families). Among the localities, Bandra (W) (61%) and among communities, Bohras (51.3%) have the highest share of families saving money in banks.

Asset Ownership

Ration card is an asset/document that assures entitlement of individuals and families to various government provisions and welfare programmes. In Mumbai, it is considered a vital document of residential proof, and the holders of ration card are treated as a class different from the others who do not possess the same. Hence, one finds individuals and families keen to secure ration card while the local authorities appear to make it difficult to acquire. Despite one's entitlement to a ration card, it is not uncommon to find officials asking for bribes or playing tricks to avoid providing the same. Given that Muslims are treated with contempt and that the identity of many who belong to West Bengal or northern states of India are strategically obfuscated with that of Bangladeshis, securing a ration card by Muslim families in the city gets more difficult [*cf* Committee for Protection of Democratic Rights (CPDR) and Women's Research and Action Group (WRAG) 1998]. The present study shows that a significant proportion of families, 8.2 per cent, in the city do not own ration cards. There is locality and sect/community-wise variation in the ownership of ration cards. While 97.4 per cent of the families own ration card in Dongri, the share of families owning the same in Govandi is as low as 85.6 per cent and in Dindhosi the percentage of those who own ration card is 87 per cent (Table 5.6). In general, larger share of families living in the main city have ration cards, while the share is lower in suburbs. Whereas all the surveyed families of Khojas and Bohras own ration card, only 91.8 per cent Sunni (General) and 89.8 per cent Sunni (OBC) have ration card. Given the poor economic status of the Sunnis (General as well as OBCs), one feels that entitlement to get subsidized foodgrains and housing etc., are in reality denied to the community by not providing them with the ration cards.

Only about 3 per cent of Muslims own car. None of the families in Dindoshi and only 0.7 per cent of Sunni (OBC) families fall into this category. In comparison, around 13 per cent own motorcycles or scooters. While the ownership of scooters is higher in Dongri and B-N belt, the ownership of car is higher in Bandra (W). Community-wise, Bohras, Khojas and Memons have better ownership of motorcycle and car. Luxury item, air-conditioner, is owned only by 4 per cent of the families. This ownership is also skewed across localities and communities: 25.6 per cent of the

Table 5.6: Family Asset Ownership by Locality (per cent)

Assets	*Govandi*	*Kurla*	*Bandra (E)*	*Bandra (W)*	*Dindhosi*	*Dongri*	*B-N belt*	*Total*
Ration Card	85.6	95.1	90.5	90.2	87.0	97.4	94.6	91.8
Fan	95.7	99.0	99.4	98.8	99.2	99.5	97.8	98.1
Television	62.4	79.9	81.1	89.0	78.9	91.7	88.8	79.5
Cooking Gas	52.8	69.3	87.0	65.9	67.5	97.9	92.3	74.8
Refrigerator	24.3	43.6	50.3	48.8	22.0	79.3	74.4	48.6
Computer	2.1	6.7	5.9	30.5	0.0	17.6	16.6	9.4
Motorcycle/ Scooter	4.0	8.5	10.7	17.1	4.9	29.0	21.7	12.8
Car	1.1	1.8	3.0	24.4	0.0	3.1	2.6	3.0
Telephone/ Mobile	59.5	72.9	82.2	80.5	76.4	89.1	84.3	75.5
Air-conditioner	1.1	1.0	3.6	25.6	0.0	10.9	5.4	4.4
Sewing Machine	18.4	28.9	29.6	34.1	35.0	29.5	28.1	27.2

Source: As for Table 5.1.

families in Bandra (W) own air-conditioners while none of the families in Dindhosi possesses the same. Around 23 per cent of Bohras in comparison to only one per cent of Sunni (OBC) own air-conditioners. About 80 per cent of the sample Muslim families own a television. Here too, variation across localities and communities is observed. While almost 92 per cent of the families in Dongri own a television, it is as low as 62.4 per cent in Govandi. Among Sunnis (OBC), 70 per cent own a television. About 95 per cent of the Bohra community owns a television. This questions the myth that Muslims do not watch television and consider it a 'satanic/idiot box'. These families have attempted to procure the same, though the TVs may be of low quality or from second hand/ junk market discarded by the upper and middle class.

Refrigerators are owned by 49 per cent of the families. About 79 per cent of the families in Dongri own a refrigerator, and the same is only 22 per cent in Dindhosi and 24.3 per cent in Govandi. Computer is possessed by only 9 per cent of the families, and in comparison, 75.5 per cent of the families own mobile/telephone, which can be considered a necessity for self-employment related communication. Women in Muslim families are largely acquainted with sewing and embroidery. In fact, many of the families earn their bread and butter by this. This is one of the reasons for relatively high ownership of sewing machines (27.2%) among the

Muslim families in the city. It is interesting to note that there is not much variation across localities and communities in the ownership of sewing machine as has been seen in the case of ownership of car, motorcycle, scooter, etc.

About 98 per cent of the sample Muslim population owns a fan. Interestingly, even in B-N belt (2%), Dongri (0.5%) and Bandra (W) (0.2%) all households do not own fan. The percentage for the same is 4.3 for Govandi. Approximately 75 per cent of Muslim families own cooking gas, and the possession of the same is the lowest in Govandi (52.8%). Again, higher percentage of families from Bohras, Khojas and Memons own cooking gas, while only 65.3 per cent of Sunnis (OBC) families own the same. The data demonstrate that Sunni (OBC) are the most deprived section among Muslims, followed by Sunni (General). As far as locality is concerned, Govandi is the most deprived area followed by Dindhosi.

Employment and Unemployment

In India, the employment market has traditionally been segmented and the new employment opportunities which the new economy has thrown up have also been largely determined by factors like technical knowledge, quality and level of education. At the same time, traditional discriminatory factors like religion, caste, creed, region and gender have not lost their importance in determining access to employment opportunities. As we will see below, these factors are perhaps more constraining to Muslims.

A large share of Muslim population is out of work force: 21.3 per cent housewives, 30.7 per cent students and 8.8 per cent children (Table 5.7). Thus, effectively only about 39 per cent of the population is available for work. Out of the total population, 5.5 per cent are unemployed but seeking employment, while 4.3 per cent are not available for work. The work participation rate in the age-group of 15-60 years among Muslims in the city is 42.3 per cent: 72.6 per cent for males and 9.4 per cent for females. In the age-group of 15 years and above, the total work participation rate is 42.2 per cent: 72 per cent and 9.7 per cent, respectively, for males and females. With reference to those in labour force (excluding children, housewives, students and those not available for work) in the age-group of 15-60 years, the unemployment rate among

Table 5.7: Occupation of Individuals by Sex

Occupation	*Sex (%)*			*Sex (counts)*		
	Male	*Female*	*Total*	*Male*	*Female*	*Total*
Child	8.60	8.90	8.80	422	400	822
Student	30.90	30.50	30.70	1508	1364	2872
Housewife		44.50	21.30	0	1994	1994
Unskilled and semi-skilled manual workers	9.20	2.70	6.10	447	121	568
Skilled manual workers	6.10	0.60	3.50	298	29	327
Religious services	0.20	0.10	0.10	8	3	11
Teacher (Pvt. & Govt.)	0.30	0.60	0.40	13	29	42
Govt. employee grade c & d	0.40	0.10	0.30	21	3	24
Govt. employee grade a & b	0.30	0.00	0.20	16	2	18
Executive in private companies	0.70	0.20	0.40	33	8	41
Professionals (doctor, engineers etc. not in Govt. services)	1.20	0.40	0.80	57	16	73
Other services	0.90	0.30	0.60	45	14	59
Other services—not fixed job	5.20	0.90	3.20	254	41	295
Self-employed	25.90	0.80	13.90	1263	35	1298
Unemployed but seeking employment	6.10	4.80	5.50	298	215	513
Unemployed but not seeking/available for work	4.10	4.60	4.30	200	205	405
Total	100.00	100.00	100.00	4883	4479	9362

Source: As for Table 5.1.

males population is 10.9 per cent, while the same is 43.1 per cent for the females (Table 5.8). The aggregate unemployment in this age-group is 16 per cent. The unemployment rate across all localities and communities is higher among females (Figures 5.1 and 5.2). Among localities, Govandi and Dongri, and among sects/communities, Memons and Sunnis (both General and OBCs) show relatively higher unemployment rate.

In the age-group of 15 years and above, the unemployment rate is 15.4 per cent: 10.5 per cent for males and 41.5 per cent for females. The higher employment rate specifically among women in Muslims challenges the myth that they are either not allowed to work or do not want to work (see Khan, 2007 for further details). A comparison of the present study with that of NSSO 61st round data for Mumbai shows the followings: First, in comparison to males from all religion, the employment rate among Muslim males is higher. However, Muslim females are relatively less employed than females from all religions in the city (Table 5.9). Second,

Table 5.8: Share of Worker (in age-group of 15-60 years) in Various Categories of Employment/Unemployment by Sex (per cent)

Employment Category	*Sex*		*Total*
	Male	*Female*	
Unskilled and semi-skilled manual workers	16.3	23.8	17.4
Skilled manual workers	11.1	5.9	10.3
Religious services	0.3	0.4	0.3
Teacher (Pvt. & Govt.)	0.5	5.9	1.3
Govt. employee grade c & d	0.8	0.6	0.8
Govt. employee grade a & b	0.6	0.4	0.6
Executive in private companies	1.3	1.6	1.3
Professionals (doctor, engineers etc. not in govt. services)	2.2	3.3	2.3
Other services	1.7	2.8	1.9
Other services—not fixed job	8.1	5.1	7.7
Self-employed	46.3	7.1	40.1
Unemployed but seeking employment	10.9	43.1	16.0
Total	100.0	100.0	100.0

Note: Many of the professionals are also self-employed besides working in private companies. Thus, the share of those self-employed may be slightly higher than that presented in the Table.

Source: As for Table 5.1.

Muslim males are relatively more concentrated in self-employment, while there is negligible presence of Muslim females in this category. Third, the shares of unemployed Muslim males and females are much higher than those for all the religions. Lastly, although the share of Muslim males is higher in labour force, the share of Muslim females in the same is lower. Hence, the share of Muslims in labour forces (age 15+ years) as a whole comes down below the NSSO figure of the same for all the religious groups for the city.

The employment pattern among Muslims is largely dominated by males. The two occupations that show a substantial presence of females in comparison to males are teaching and gynaecology. Muslim representation in Government services (other than teachers) in the age-group of 15-60 years is extremely poor, 1.6 per cent. The share of unskilled and semi-skilled manual workers to the total workforce (including those unemployed but seeking employment) is about 27 per cent. This shows that most of the workers from the Muslim community are located at the bottom of the pyramid. They have hardly any presence in government

Table 5.9: Distribution (per cent) of Persons of age of 15 Years and above by Broad Usual Activity Status in Mumbai

Source	*Employed*	*Self-employed*	*Unemployed*	*Not in Labour Force*
Male				
Muslims	80.4	37.1	8.5	19.6
NSSO 61st Round (all religion) 2004-05*	78.6	31.7	2.5	18.9
Female				
Muslims	16.5	1.1	6.9	83.5
NSSO 61st Round (all religion) 2004-05*	26.7	7.8	1.5	71.8
Total				
Muslims	49.9	19.9	7.7	50.1
NSSO 61st Round (all religion) 2004-05*	54.5	20.6	2.0	43.5

Sources: 1. * NSSO (2007).
2. The rest of the estimates are based on the household survey.

services and their share with respect to their population in executives and professionals is also quite small.

Despite the rampant poverty, the share of child workers (to the total population) among Muslims is small. About 0.7 per cent

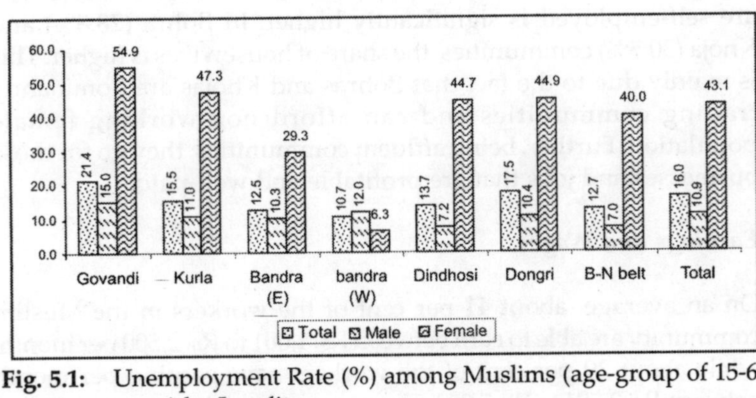

Fig. 5.1: Unemployment Rate (%) among Muslims (age-group of 15-60 years) by Locality.

Source: As for Table 5.1.

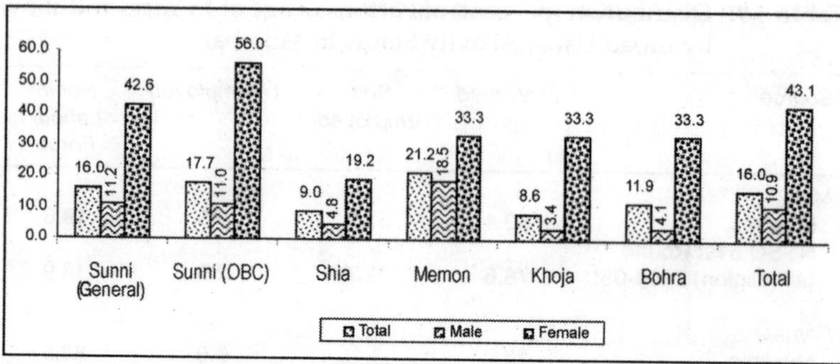

Fig. 5.2: Unemployment Rate (%) among Muslims (age-group of 15-60 years) by Sect/Communities.

Source: As for Table 5.1.

of the total population in the age-group of 7-14 years claim to be seeking employment and only about 0.2 per cent are working (Table 5.10). The share of those self-employed is highest among the age-group of 35-59 years and it is about 18 per cent in the age-group of above 60 years. The share of unemployed population is higher in the lower age-group (15-34 years). The unsuccessful struggle for formal employment in this age-group perhaps results in shift to self-employment in latter age-group, in which the share of self-employed swells. Locality-wise, there is not much variation in the pattern of employment. However, in Dindhosi (17.9%), Dongri (15.3%) and Bandra (W) (14.3%), the share of those who are self-employed is significantly higher. In Bohra (28.8%) and Khoja (30.5%) communities, the share of housewives is higher. This is mainly due to the fact that Bohras and Khojas are dominantly trading communities and can afford non-working female population. Further, being affluent communities, they go for those businesses and jobs that are profitable and well-paid.

Earnings and Wages

On an average, about 41 per cent of the workers in the Muslim community are able to earn between Rs. 1,001 to Rs. 2,500 per month, while about 39 per cent of the workers have earning per month between Rs. 2,501 to Rs. 5,000. The proportion of those earning below Rs. 1,000 per month is only about 3.7 per cent. Out of those who work for religious services, 27.3 per cent have no income. Only about

Table 5.10: Occupation of Individuals with Reference to Age (per cent)

Occupation	*Age Category (years)*						
	Below 6	*7-14*	*15-24*	*25-34*	*35-59*	*Above 60*	*% of Total Population*
Child	62.7	3.6					8.8
Student	37.3	94.7	37.5	1.5			30.7
Housewife		0.1	13.4	39.2	41.4	34.3	21.3
Unskilled and semi-skilled manual workers		0.1	5.6	9.5	11.6	6.9	6.1
Skilled manual workers		0.1	4.1	5.8	6.1	1.5	3.5
Religious services			0.1	0.2	0.2	0.2	0.1
Teacher (Pvt. & Govt.)			0.4	0.8	0.9		0.4
Govt. employee grade c & d			0.1	0.3	0.8		0.3
Govt. employee grade a & b			0.0		0.8		0.2
Executive in private companies			0.7	0.9	0.5		0.4
Professionals (doctor, engineers etc. not in govt. services)			0.5	2.7	0.9		0.8
Other services			0.8	1.2	0.9	0.2	0.6
Other services—not fixed job			3.4	5.2	2.8	15.7	3.2
Self-employed			10.2	23.0	28.1	17.8	13.9
Unemployed but seeking employment		0.7	14.5	7.8	2.2	1.0	5.5
Unemployed but not seeking/available for work		0.6	8.6	1.9	2.8	22.4	4.3
Total (persons)	100.0 (N=1217)	100.0 (N=1630)	100.0 (N=2268)	100.0 (N=1532)	100.0 (N=2237)	100.0 (N=478)	100.0 (N=9362)

Source: As for Table 5.1.

4 per cent of the total workers in the community are able to earn an income between Rs. 10,001 to Rs. 20,000, while 2 per cent and 0.3 per cent have monthly earnings between Rs. 20,001 to Rs. 50,000, and Rs. 50,001 and above, respectively.

The mean monthly earning of the employed in different occupations ranges between Rs. 2,509 (unskilled and semi-skilled manual workers) per worker to Rs. 22,944 (in case of government grade a & b employees). The larger share of workers is engaged in self-employment where the mean earning is Rs. 4,265 per worker per month. The mean monthly income of professionals and executives in private companies is also not very high: Rs. 14,714 per month per worker in case of professionals and Rs. 13,463 in the case of executives in private companies. Thus, the above-mentioned categories of jobs are misnomers as far as earnings are concerned. There is a great deal of variation in earning between male and female workers. In almost all the employment categories (except 'other services—not fixed job'), women earn less than males. Even in the case of A & B grade Government employment, the mean income of females is found to be substantially lower than that of the males. This might be due to concentration of females in the lower rung of the respective Government services indicating stagnation in their careers. While regressing monthly earnings (*Y*) on age, sex (categorical variable, female = 1), and years of schooling, one finds that sex is the major determinant of the monthly earning: on an average females have mean monthly earning Rs. 1,819 less than males. Years of schooling (*schooling*) is another major determinant. On an average, one year of additional schooling raises the monthly earning by Rs. 280. Age, though statistically significant, is not a major determinant of earning. Age-wise classification of mean earnings shows that in most of the job categories, mean income is higher for the age-group of 35 to 59 years. However, young professionals (age-group of 15-34 years) and those in the 'other services' are able to earn more than their seniors. This is due to the fact that the new economy requires new set of skills which some of the younger ones are able to acquire and so earn higher. The equation suggests that among others, increase in educational level of the community could be a significant way forward for enhancing earnings. Further, the traditional female bias in occupation and wage differentials need to be overcome.

$$Y = 607.87 + 57.55\,age - 1818.72\,sex + 279.74\,schooling \quad ...(1)$$
$$t = (1.98)^{*} \quad (12.96)^{**} \quad (-12.49)^{**} \quad (14.02)^{**}$$

$R^2 = 0.245$; F= 181.10**

where = * significant at 5 per cent level, and
** significant at 1 per cent level of significance.

The earnings of workers across localities also show significant variation. While an unskilled and semi-skilled manual worker is able to earn Rs. 2,774 per month in Govandi, the level of earning in the same category of worker in Dindhosi is only Rs. 1,995. In case of skilled manual workers, the highest earning (Rs. 5,475 per month) is in the B-N belt, while the teachers are better paid (Rs. 15,000 per month) in Bandra (W). As expected the earnings of professionals is higher in Bandra (W) (Rs. 32,500 per month) and those in 'other services—not fixed job' also earn higher (Rs. 24,246 per month) in this rich class area. Using location dummes, equation (2) for monthly earning of workers has been estimated. The equation also shows that mean earnings of workers in Bandra (W) is significantly higher than that in Govandi (to avoid dummy trap, dummy variable for Govandi is avoided), and there is no significant differences among mean earning between Govandi and other locations.

$$Y = 588.82 + 55.96\,age - 1821.18\,sex + 268.73\,schooling$$
$$t = (1.82) \quad (12.61)^{**} \quad (-12.61)^{**} \quad (13.24)^{**}$$

$$-66.47.83\,Kurla - 82.26\,Bandra\,(E) + 4164.59\,Bandra\,(W) + 63.95\,Dindhosi$$
$$(-0.32) \quad (-0.34) \quad (11.17)^{**} \quad (-0.22)$$

$$-153.95\,Dongri - 41.84\,B\text{-}N\,belt$$
$$(-0.57) \quad (-0.18)$$

$R^2 = 0.275$; $F = 77.65^{***}$

where = * significant at 5 per cent level, and
** significant at 1 per cent level of significance.

It is a well known fact that education has an impact on employment. Type of employment and employability is largely related to the educational attainment of individuals. This is reflected in the data on employment and educational attainment presented in Table 5.11. A huge proportion of illiterates are unemployed but seeking job (11.4%), and the equal proportion is in unskilled and semi-skilled manual jobs. Further, about 46 per cent of the illiterate are concentrated in housewife segment. The large share of illiterate (also true for other education categories), 16.6 per cent, is also self-employed: most of them are hawkers or deal with waste metal, paper and scraps. Some of them are also vegetable and fruit vendors.

Literate without formal education have an employment pattern similar to that of illiterate. However, those below primary education are mainly students, housewives and self-employed. As we move towards higher education, with the decline of student population, the share of those engaged in self-employment rises. About 22 per cent of those who are educated up to secondary level are self-employed, while the proportion of self-employed among graduates is as high as about 21 per cent. The percentage of unemployed person ranges from 5 per cent to 6.5 per cent for those educated up to middle to graduation levels.

The higher percentage of self-employment and unemployment among the educated is the main reason for many Muslims not realizing advantages of education especially higher education. If the absorption of the educated in services (both government and private) is increased, the catch up syndrome may develop and many Muslims might take higher education in more positive manner. However, for this the government has to create an atmosphere where the Muslims are not discriminated in selection process and additionally might be given some preference through positive discrimination or reservation/quota. Further, as a major share of workers is engaged in self-employment, the provision of timely and easy credit and training in relevant field becomes very significant. At present, it is found that only a small proportion of self-employed families are able to utilize the institutional credit, which adversely affects the growth of their business, keeping their income at subsistence level. For many of those who succeed in getting above the poverty line, the economic condition remains so fragile as to push them back to poverty even with smallest of incidence requiring money. This incident might refer to as non-negotiable a matter as health.

Table 5.11: Occupational Category by Educational Level (per cent)

Occupation	*Level of Education*												
	Illiterate	*Literate without Formal Education*	*Below Primary*	*Primary*	*Middle*	*Secondary*	*Higher secondary*	*Graduation*	*Engg./ Technology*	*Medicine*	*Other Professional Courses*	*Child*	*Total*
(1)	(2)	(3)	(4)	(5)	(6)	(7)	(8)	(9)	(10)	(11)	(12)	(13)	(14)
Child												100.0	8.8
Student			80.9	70.5	35.9	17.0	33.0	19.2	35.3	42.3	6.0		30.7
Housewife	45.7	59.5	7.8	11.4	22.0	26.9	21.7	17.0	5.9		22.0		21.3
Unskilled and semi-skilled manual workers	11.4	8.1	1.4	2.3	7.1	9.1	5.4	6.3			2.0		6.1
Skilled manual workers	3.7	1.4	2.0	2.4	4.4	5.4	4.1	1.9	3.9		4.0		3.5
Religious services	.1	1.4		.2		.1	.2	.2			2.0		.1
Teacher (Pvt. & Govt.)						.4	.8	4.8			6.0		.4
Govt. employee grade c & d					.2	.3	.7	1.1			2.0		.3
Govt. employee grade a & b							1.1	1.0			4.0		.2
Executive in private companies						.4	.9	4.4	2.0		2.0		.4

(Contd.)

(*Contd.*)

(1)	(2)	(3)	(4)	(5)	(6)	(7)	(8)	(9)	(10)	(11)	(12)	(13)	(14)
Professionals (Doctor, Engineers etc. not in Govt. services)			.2			.1	1.6	3.6	39.2	50.0	4.0		.8
Other services	.1	1.4			.4	.6	1.5	4.0			4.0		.6
Other services — not fixed job	4.4	5.4	1.4	1.3	3.0	4.3	4.0	6.1	2.0		12.0		3.2
Self-employed	16.6	12.2	3.9	7.6	14.8	22.0	16.4	20.8	7.8	3.8	16.0		13.9
Unemployed but seeking employment	11.4	4.1	1.2	2.5	6.6	6.3	6.2	5.1	2.0	3.8	4.0		5.5
Unemployed but not seeking/available for work	9.6	6.8	1.1	1.8	5.5	7.2	2.5	4.6	2.0		10.0		4.3
Total	100	100	100	100	100	100	100	100	100	100	100	100	100

Source: As for Table 5.1.

Housing Situation and Living Environment

Urban societies and spaces in India are undergoing amazing arrangements and transformations. The city spaces are being subtly reorganized on the basis of caste, class, region and religion. The seed of religious acrimony which was sown in mid-1980s and early 1990s, is showing its effect and dividing the citizens living together for hundreds of years. The religious division of urban spaces is extremely visible on the cityscape of Mumbai. Areas like, Dongri, Nagpada, Byculla, Dockyard, Bandra (East), Kurla, Jogeshwari, Sonapur, Govandi, Zakaria Nagar (Mankhurd-Shivaji Nagar), and Cheeta Camp, are typical Muslim concentration areas/ghettoes (Shaban, 2008). This ghettoisation of Muslims is mainly because of (a) fear of life-security during riots, (b) increased discrimination against Muslims in housing market, (c) adverse economic situation and (d) largely non-availability of institutional credit to buy houses in middle and high class areas, generally having better infrastructural facilities. Here it is important to note that to some extent, the 'community feeling' and desire to live together also contributes to the ghettoisation of Muslim community in the city.

The survey data shows that around 21 per cent of the Muslim population lives in slums, while another 70 per cent reside in the low class areas of the city (based on visual appearance of the area and house types). Only about 5 per cent are located in middle class areas and 1.6 per cent in high class areas. The economic hardship is one of the main reasons for undue concentration in low class and slum areas. The proportion of the Muslim households living in class-wise mixed areas is about 2.6 per cent. Among the localities chosen for the survey, in Bandra (W) about 44 per cent of the families live in slums. Bandra (W) is followed by Govandi (37.9%), Kurla (25.3%), Bandra (E) (19.5%), and B-N belt (9.3%). It should be noted that as most of the samples in Dindhosi were taken from MMRDA resettlement colony, we do not have any representation of slum households from this area. In Dongri, only 0.5 per cent of the families are located in slums. Out of the total surveyed families (1643) from the seven major areas, about 74 per cent live in ghettoes. Although it ensures a relative physical safety, the concentration leads to many adverse socio-economic consequences. The ghettoisation effectively cuts off the Muslims from the mainstream population, which among other, results in lower educational attainment, conservative attitude and lower employment rates. This necessitates formulation and implementation of effective policies to make housing for Muslims in mixed areas, higher and middle class residential colonies, more accessible and safe.

House Types and Ownership

Only about 31 per cent of the Muslim families in the city are living in flats (Table 5.12). This shows that the Muslim community is going through severe housing problem and marginalization in the city. Locality-wise analysis shows that as much as 84 per cent of the families in Kurla live in slums/chawls, while the same in Govandi is about 68 per cent. In general in all the areas, except in Dindhosi (where most of the surveyed families live in MMRDA provided flats) and Bandra, most of the families from the community are located in poor housing conditions and in areas having inadequate infrastructural facilities. Further, except Bohra families, more than two-thirds of families from all other communities are located in chawls/slums. Among the Bohras (46.2%), Sunni (General) (33.0%) and Memons (33.3%), the ownership of flats is relatively higher in comparison to other communities.

Table 5.12: Distribution of Sampled Families as per Locality and Type of House (per cent)

Locality	*Type of House*			*Total*
	Flats	*Hut*	*Chawl*	
Govandi	18.7	13.1	68.3	100
Kurla	13.2	3.4	83.5	100
Bandra (E)	50.9	1.2	47.9	100
Bandra (W)	42.7		57.3	100
Dindhosi	99.2	.8		100
Dongri	23.9	1.0	75.1	100
B-N belt	29.7	3.2	67.1	100
% of Total Families	30.6	4.7	64.7	100

Source: As for Table 5.1.

The ownership (the ownership of houses reported is based on the claim of the respondents across hutments to flats) of houses by Muslim families ranges from about 74 per cent in Govandi to 90 per cent in Dongri. In Govandi, a large number of families live in rented rooms. The reason for this lies in the availability of low cost rooms in this area. However, these rooms are often overcrowded. Overall, the ownership of houses by the families in the city is about 82 per cent. There is much variation in community-wise ownership of the houses: a significant percentage of Khojas own their residence (91.7%), while the least ownership of houses is found in the case of Shias (76.1%).

Room-Density

As has been mentioned earlier, Muslims in the city are living in adverse environmental and housing conditions. In more than three-fourths of the total surveyed families, the average room density is more than 3 persons per room (Table 5.13). The situation is worse in Bandra (E) and Dindhosi where, in about 41 per cent and 47 per cent of the surveyed families, respectively, 5-10 persons live in a room. This overcrowding is an important factor behind lower educational attainment in the community. Sect/Community-wise there is not much difference in room density except that Bohras and Khojas are relatively better off. This gives an idea of the stress and strain faced by the community especially the youth and married couples due to the lack of availability of private space.

Facilities within House

The unsatisfactory living environment of the Muslim families is further proved by the fact that most of the families do not have

Table 5.13: Distribution of Number of Persons per Room by Locality (per cent)

Persons per Room	*Locality*							*Total*
	Govandi	*Kurla*	*Bandra (E)*	*Bandra (W)*	*Dindhosi*	*Dongri*	*B-N belt*	
Below 1	2.7	.3	3.0	4.9		1.6	3.5	2.1
1 to 3	23.2	21.6	11.2	35.4	17.9	30.1	25.9	23.1
3 to 5	32.5	38.4	41.4	32.9	33.3	45.6	43.8	38.6
5 to 10	37.9	37.1	40.8	24.4	47.2	22.3	26.5	34.0
Above 10	3.7	2.6	3.6	2.4	1.6	.5	.3	2.2
Total	100.0	100.0	100.0	100.0	100.0	100.0	100.0	100.0

Source: As for Table 5.1.

toilets in their houses (Figure 5.3). A large share of families has to depend either on public facilities/community toilets, or defecate in open spaces. Out of the total surveyed families, only about 40 per cent have a toilet in their houses. In areas like Kurla, Dongri and Govandi, the situation is worse. Further, with reference to community/sect, there is hardly any variation in terms of ownership of above-mentioned facilities, except that Bohra community is slightly better in ownership of bathrooms and toilets.

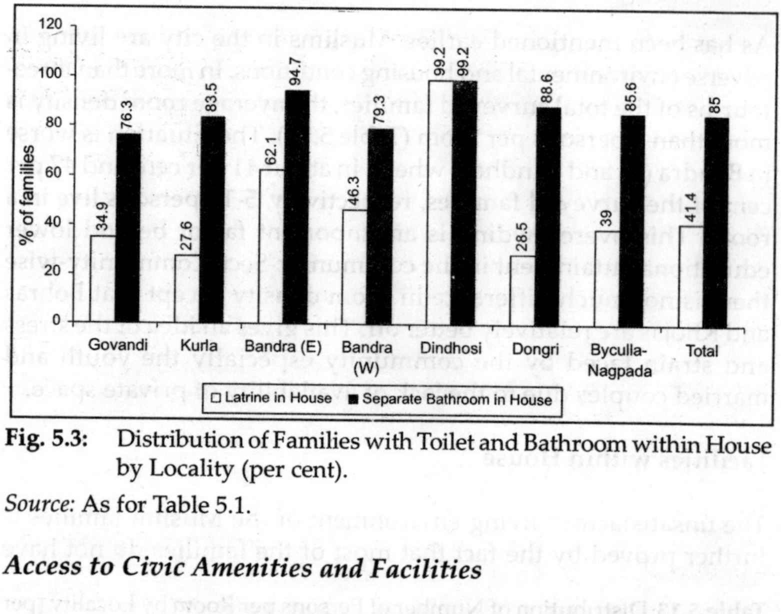

Fig. 5.3: Distribution of Families with Toilet and Bathroom within House by Locality (per cent).

Source: As for Table 5.1.

Access to Civic Amenities and Facilities

Access to civic amenities by the Muslim community has been measured by assessing access to water and electricity. The main source of drinking water across the localities in the city remains municipal tap water. In Govandi about 23 per cent of the families do not get municipal tap water. Only in Dindhosi and Bandra (W), the supply of water is adequate throughout the year. All the other areas face major problems, the worst being Govandi and Kurla, where around 47 per cent and 37 per cent, respectively, of the total families face water shortage throughout the year. Even in areas like Govandi and Kurla, a significant proportion of the families depend on water purchased from tankers or the local informal water market. This makes a significant dent on the earnings of these marginalised families and constrains their saving and investments in productive activities. Often one has an impression that in a city like Mumbai all families have access to electricity. However, the survey shows that among Muslims, about 3 per cent of the families have no electricity connection and a majority of such families are living in Govandi.

Conclusion

A majority of Muslim families in the city live in penury and deprivation. About one-third of the families live below poverty line, but about two-thirds are extremely vulnerable in terms of income. Their access to institutional credit is extremely poor, only about 1.9 per cent of the families in the city have credit from banks and co-operatives. Most of the areas where Muslims live are informally designated as 'negative' areas by the banks; therefore, access to credit to Muslim families is limited. The asset ownership also paints a bleak picture. It is surprising that even ration card is not owned by a significant proportion of families in areas like Govandi and Dindhosi. A substantial share of workers among Muslims in the city is self-employed. The share of the Muslim population in the Government and private sectors is nowhere near the proportion of their population. It will not be an exaggeration to state that even in urban areas, Muslims remain in subsistence occupations, that too in a period when the salaries are sky-rocketing and wages are growing much faster than the national income. Muslims are largely concentrated in slums and lower class areas in the city. The ghettoized living has become a way of life for Muslims. Riots have been instrumental in creating a socio-cultural divide and minority groups like Muslims are forced to live in ghettos for their economic and biological security. Further, discrimination in housing market and limited access to financial services from formal institutions for buying houses also keeps a majority of them in the low class and slum areas. To alter the situation multi-pronged and dedicated efforts from the Government are called for. The need of the hour is to enhance credit facilities to the community and provide training and opportunities for growth to those who are engaged in self-employment. Further, the Muslim community needs an internal reform to encourage work participation of women. The

Government needs to come out with programmes and policies which ends discrimination against Muslims in jobs (both, in the private and public sectors). In fact, the moral and economic upliftment of the Muslims demand similar actions from the government as has been for the SCs and STs in the country.

[**Acknowledgement**: The author gratefully acknowledges the financial help and cooperation received from the Maharashtra State Minorities Commission, Government of Maharashtra, Mumbai, for the study. The help received from Ranu Jain at various stages of the study are also gratefully acknowledged.]

REFERENCES

CPDR and WRAG (1998). "Deportations of Bengali-Speaking Muslims from Mumbai: A Fact Finding Report", Mumbai, August 27.

Government of India. District Profile: Mumbai and Mumbai Suburban, *Census 2001*.

Government of India (2006). "Social, Economic and Educational Status of the Muslim Community in India", Prime Minister's High Level Committee, Cabinet Secretariat, Government of India, New Delhi, November.

Khan, Sameera (2007). "Negotiating the Mohalla: Exclusion, Identity and Muslim Women in Mumbai". *Economic and Political Weekly*, 42(17): 1527-33.

Masselos, Jim (1994). "Postmodern Bombay: Fractured Discourses" in Sophie Watson and Kathie Gibson (eds.), *Postmodern Cities and Spaces*, (Oxford: Basil Blackwell, Oxford) 199-215.

National Sample Survey Organisation (2007). *Employment and Unemployment Situation among Major Religious Groups in India 2004-2005, NSS 61st Round (July 2004 - June 2005)*. Ministry of Statistics and Programme Implementation Government of India, New Delhi, March.

Phadke, Shilpa (2007). "Dangerous Liaisons: Women and Man—Risk and Reputation in Mumbai", *Economic and Political Weekly*, 42(17): 1510-18.

Robinson, Rowena (2005). *Tremors of Violence: Muslim Survivors of Ethnic Strife in Western India*, New Delhi: Sage Publications.

Shaban, Abdul (2008). "Ghettoisation, Crime and Punishment in Mumbai", *Economic and Political Weekly*, 42(23): 68-73.

Sharma, R.N., K. Sita, M. Vyas, M. Jha and A. Shaban, (2008): "Impact Assessment of Resettlement Implementation for Mumbai Urban

Transport Project (MUTP)-II", Research Report, Tata Institute of Social Sciences, Deonar, Mumbai.

Times of India (2008): "Urban Poverty in Maharashtra Jumps to 18%", Mumbai, January 24.

6

Encountering Socio-spatial Exclusion: The Experiences of Muslims in Kolkata

Sohel Firdos

Using household level data generated by a survey in Kolkata, this chapter deploys 'socio-spatial exclusion' framework to understand the existing deprivation and marginality of Muslims in the city. First, since there is an overlap between slum areas and the areas inhabited by Muslims, this chapter has attempted to find out the socio-economic status of slum dwellers and the deficiency of public goods and services provided to them. Second, it analyses the assets available with them to assess the vulnerability of such people in a situation of stress. Third, it examines how the public policies are leading to the relative deprivation of Muslims in terms of their access to a service like education. Finally, it probes the processes through which the new urban economies of the city are excluding a majority of the Muslims from employment because of their association with certain traditional occupations such as bakery, garment and kite making. Taking note of Sen's (2000) argument that to be excluded from common facilities or benefits that others have can certainly be a significant handicap that impoverishes the lives that individuals can enjoy, this essay talks about deprivations of Muslims living in low income areas, primarily slums of Kolkata, where social and spatial exclusion converge. It is argued that the disadvantages experienced by such people are the products of exclusions from shared opportunities such as basic municipal services of the city and educational facilities.

Theoretical Perspectives

The concept of social exclusion has made valuable contributions to the current understanding of poverty and socio-spatial polarisation. The findings of a study commissioned by International Labour Organisation (ILO) on social exclusion suggest that the concept of social exclusion has certain advantages. It provides an integrated and dynamic analytical perspective which reveals the processes, agency, and multi-dimensionality of disadvantage. It provides a framework for analysing the relationship between livelihood, well-being and rights. And by relating social and economic status to the notion of citizenship, it generates a collective moral responsibility for social integration. However, it is pertinent to mention here that despite wider application of social exclusion framework in understanding deprivation, there has been considerable debate in defining social exclusion. To Rodgers, it is basically about inadequacies with respect to certain means of living. Social exclusion is seen in the growth of homelessness or urban slums, the declining hopes of the long-term unemployed, the lack of access to jobs and incomes of migrants and some ethnic minorities, the increasingly precarious nature of jobs on offer to new labour market entrants (Rodgers, 1995).

De Haan (1995) finds that the potential of the approach for understanding poverty lies in the fact that present poverty studies tend to emphasize economic aspects and to pay less attention to political and cultural dimensions of poverty. The analysis of the inter-related processes of inclusion and exclusion could help one to understand the interplay between these dimensions. Nayak (1994) identifies the following types of exclusion: exclusion due to poverty, exclusion from employment and means to livelihood and caste based exclusion. Social exclusion is a process which restricts the access of certain social groups to valued resources and entitlements, relegating them to the status of social outsiders. It is a multi-dimensional concept, conceived to capture different forms of social disadvantage — economic, social, political and cultural — that persist, in multiple variants and with different intensity, across nations. Consequently, in its widest connotation, the concept encompasses theories of poverty, inequality, racial/ethnic cleavages, long-term unemployment and citizenship (Appasamy *et al.*, 1995). Amartya

Sen (2000) argues that social exclusion helps to understand poverty better and in a more comprehensive way. While the underlying idea behind the concept of social exclusion is not radically new, the growing literature on the subject has helped to enrich causal understanding and empirical analysis of certain aspects of poverty and deprivation. To be excluded from common facilities or benefits that others have can certainly be a significant handicap that impoverishes the lives that individuals can enjoy. No concept of poverty can be satisfactory if it does not take adequate note of the disadvantages that arise from being excluded from shared opportunities enjoyed by others.

Entitlement and Social Exclusion

A useful way of conceptualizing the system of economic and social exclusion would be to focus on the problem as being one of lack of entitlement of economic and social power amongst a large section of the population. Following the pioneering formulation of Sen (1981) in the context of famines, the notion of entitlement suggests that it is not a drop in total food production but its poor distribution that may lead to famines. In fact, there may well be famines in years of normal or even above normal food grain production if the distribution of purchasing power or entitlement is so skewed as to leave a large segment of the population below the margin of subsistence. The notion of entitlement refers to the actual or effective empowerment of a person to trade his original endowment of labour power and other factor incomes for food and other basic necessities. Thus, it is not the total production of basic necessities that would make a material difference to the inherent health or strength of a community so much as the distribution of this total amongst the population. Thus, the concept of the entitlement lends itself rather appropriately to address the question of social and economic exclusion in a wider context.

There is a fundamental sense in which the notion of entitlement is rooted in the neo-classical tradition of economics. The ability of individual agents to purchase goods and services is defined by their endowments, in the form of their labour power and other factor incomes and the exchange opportunities that are available. If it is the original endowment, or the distribution of purchasing power that is skewed then, quite clearly, the resultant distribution of goods and services traded would also, in general, be skewed. In order to bring about an alteration in the distribution of the original distribution of factor incomes itself, one needs to look at a theory of social or group or class action that ensures a more equitable distribution in the first place or better terms of trade or both. It is this that is required to ensure an equitable distribution of economic and social empowerment amongst the widest sections of the population. One should, however, note here that entitlements are not just a question of purchasing power as there may be means of acquisition other than *via* market exchange.

Entitlements are essential because they enable individuals to gain command over goods and services, and particularly those that may be thought of as "basic" goods. Once they are armed with this, individuals are then able to enhance their capabilities to realize their full potential. A larger entitlement or purchasing power allows a wider range of choices to an individual. The phenomenon of social exclusion may then be understood as one where there are sections of the population whose entitlements are limited or negligible.

It may here be important to think of entitlements as not just economic entitlements, but also political and social entitlements. In a privilege-laden or feudal society, the latter forms of entitlements are denied to all except a small minority of elite groups. Thus, it is only an equitable distribution of economic, social and political entitlement that may be said to have the ingredients of a just society.

Social exclusion manifests itself in diverse ways. All those who are poor are necessarily part of the socially excluded in a fundamental sense, though the latter constitutes a wider category. Unemployment is, however, one of the mechanisms of exclusion. Poverty is an end state which itself implies a lack of entitlement to the basics for survival. Unemployment is a mechanism by which an adult member of society is denied a means to his or her livelihood and hence is prevented from his or her entitlement to basic goods and services. Even if the ranks of the unemployed may be compensated by a system of doles so as to enable them to acquire the basic goods necessary for their sustenance, they would still constitute an important category of the socially excluded, for employment is a vital constituent of an individual's dignity and self-respect in society.

One may identify several other important dimensions of exclusion. One may, for example, consider exclusion from education, exclusion from housing, exclusion from property ownership, exclusion from democratic participation, exclusion from access to health services, exclusion from public goods, to name but a few. In addition to these, one may think in terms of gender-based exclusion, exclusion of the old and infirm, exclusion of widows, and exclusion of the physically handicapped. Each of these deals with a special form of exclusion and deserves exploration for its own sake.

The notion of exclusion may also be widened to incorporate the notion of rights. For example, one may examine the notion of exclusion from the right to free speech, religion or, in the case of the young, the right to have a secure childhood. Exclusion from these rights constitutes a fundamental violation of individual liberty.

Finally, we may also look at a dimension of social exclusion that is of central importance in India. This has to do with exclusion on the basis of caste and religion. Caste in some sense is the most potent form of exclusion because this sociological characteristic is fixed at birth and is hence completely inflexible. Similarly, if Muslims as individuals or social groups are excluded from livelihood opportunities or public goods and services either through government actions or actors of governance, questions of citizenship can be raised. Though daunting in themselves, the problems of poverty, unemployment, illiteracy and poor access to health care may still be addressed by concerted public action by government as well as non-governmental organizations. Eradication of caste-based or religion based exclusion is, however, a much slower and yet a more daunting task.

State and the Spatial Exclusion of Muslims in Cities

Most attention is paid to exclusion as a function of excluded people and communities are excluded because of, for example, poverty, disease, disability, age, gender, sexual preference, race, religion and so on, rather than to exclusion as a function of other actors, particularly state. The best example of exclusion as a function of state may be illustrated by state sponsored violence against Muslims in Ahmedabad city. Mahadevia (2007) narrates the impact of a series of communal violence on the spatial redistribution of population in that city on religious lines, particularly the violence that took place in 2002 against Muslims perpetuated by Hindu rightist groups such as RSS and Vishwa Hindu Parishad and supported by the State Government of Gujarat.

> The intensity of the divisions between the two communities is so large that there are now bordered cities within Ahmedabad. ...the city has been divided into two cities, one, that of the Muslims and the other, that of the Hindus. The Muslim city is in several parts and hence there are four small Muslim cities and a large Hindu city. .....Geographically, the Muslim population of the city has been pushed to live in areas outside the Municipal Corporation limits [of Ahmedabad], with the exception of the Shah Alam area which is within the Ahmedabad Municipal Corporation (AMC) limits. .... Economic interests are also inter-mixed in this process of marginalizing the Muslims and pushing them into their own cities. ... (Mahadevia, 2007: 379).

Anjana P. Desai's (2002) work on Ahmedabad suggests the adverse effects of communal riots between Hindus and Muslims on the spatial segregation in the city. All the ninety families that she interviewed preferred to stay near their own religious community. A large majority of the people interviewed felt strongly that segregated neighbourhood is necessary to avoid riots as they have lost property and family members in the mixed community.

A similar kind of experience in terms of spatial exclusion of Muslims is reported from Kolkata though so far physical violence has not been used as an instrument to reorganize the city space (Firdos, 2008). There are three distinct areas in Kolkata Municipal Corporation (KMC), where a majority of Muslims live, such as Park Circus area in central part of the city, Cossipore-Chitpur area in the north and Metiabruz on the western part of the city. A large majority of the Muslims live in slums. All these three areas also specialize in production of one commodity or the other, such as most of the tanneries are/were located in Park Circus area, Cossipore-Chitpur area is known for wholesale or/and retail trade of a number of commodities starting from dry fruits to plastic containers while Metiabruz specializes in ready to wear garments. With the closure of tanneries in Park Circus area due to a court order, a large number of Muslims who used to work in the tanneries have been rendered jobless. The tanneries are supposed to be shifted in Bantala area which is far-off from their existing locations in Park Circus. Since the workers of tanneries live in the slums of Park Circus and located near to their work places, it is very difficult for them to travel to new location of tanneries in Bantala as it will take longer time for them to travel and it will also add transport cost. Many of the Muslims consider the closure of tanneries as a conspiracy to dislocate them from the central parts of the city in terms of their residence and livelihood. Though this conspiracy theory may be contested because the tanneries have been closed on the basis of court order for environmental concerns and the State Government has to act on the order, the fears of Muslims are well founded.

Demographic Characteristics of Muslims in Kolkata

Kolkata city has a sizeable Muslim population. With a population of 9,26,769, Muslims constitute 20.27 per cent of the total population of Kolkata Municipal Corporation (KMC). Muslims have a lower sex ratio, of 739 females per thousand males as against 852 among the Hindus. There is a significant difference in the literacy rates of Muslims and Hindus. The literacy rate among the Muslims is 60.09 as against 77.38 among the Hindus. Moreover, the Muslims are unevenly distributed over the city space of Kolkata and consequently, there are certain localities, namely, Tangra, Tiljala, Narkeldanga and Park Circus in the central part of the city; Metiabruz, Garden Reach and West Port areas in the north-west; and Cossipore, Chitpore area in the north, where most of their population is concentrated. One of the very striking characteristics of such localities is that a very large majority of people live in slums. Some of the municipal wards in these areas such as Ward no. 58 Tangra, Tiljala, Ward 29-Narkeldanga in the Park Circus cluster; Ward 137-Metiabruz, Ward 134-Garden Reach, West Port in Metiabruz cluster; and Ward 6-Cossipore in North Kolkata may be referred to as 'slum wards', as the entire population of these wards lives in slums. There is a complete convergence between Muslim localities and the areas with a very high slum population suggesting that a vast majority of them in the city lives in slums. Some of the slums in these areas are more than 100 years old testifying the embeddedness of the community in the low income areas without witnessing any horizontal or vertical mobility. Slums are also the most visible expression of urban poverty in the cities. Today's slums reveal the spatial dimension of contemporary urban stratification which results in growing economic, social, cultural, political and environmental inequalities. The most usual picture of slums reveals deprivation, unemployment, insecurity, broken families, social exclusion, disease, crime and violence.

Since there is hardly any data on provisions of municipal services such as drinking water and lighting provided to the Muslim households, we will examine the provisions of such services in the slum wards as a whole that may be indicative of such services provided to Muslim households. The availability of assets with the people living in these areas may be another indicator to assess the socio-economic conditions of a large number of Muslim households.

Provisions of Municipal Services

We have measured the deficiencies in municipal services by taking into consideration facilities of drinking water and sources of

Table 6.1: Population Characteristics of Selected Slums in Kolkata, 2001

Cluster	*Total Population*	*Slum Population*	*Percentage of Slum Population*
Park Circus Cluster			
Ward 58 – Tangra, Tiljala	86,618	86,605	99.98
Ward 29 – Narkeldanga	46,814	46,251	98.80
Metiabruz Cluster			
Ward 137 – Metiabruz	20,041	19,710	98.35
Ward 134 – Garden Reach, W. Port	36,625	35,836	97.85
North Kolkata Cluster			
Ward 6 – Cossipore, Chitpore	47,407	38,910	82.08

Source: Census of India 2001.

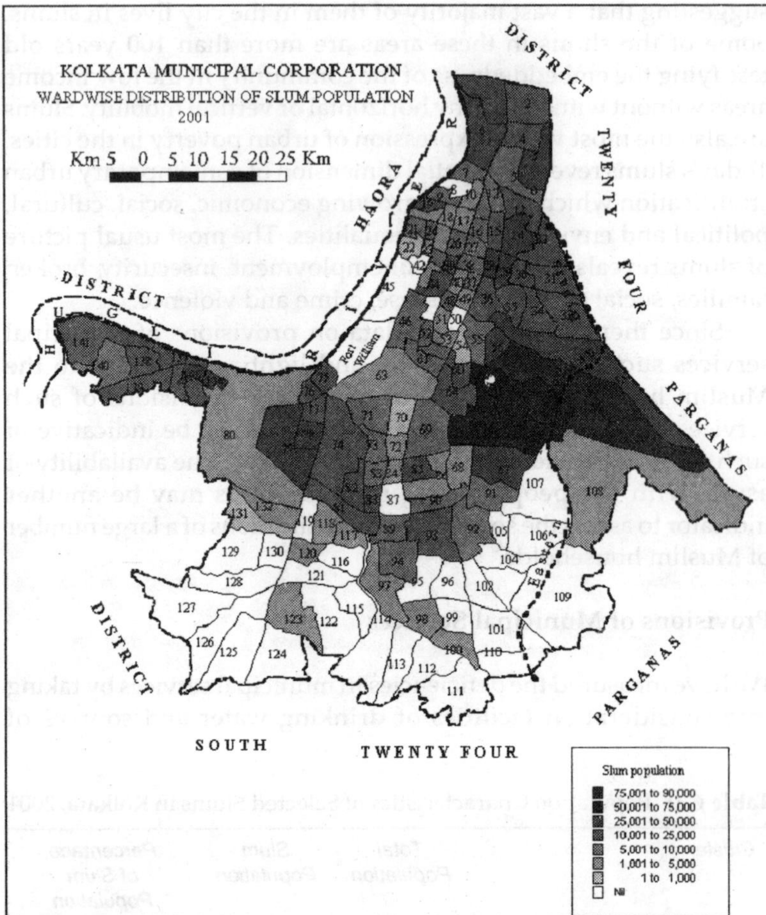

Fig. 6.1: Distribution of Slum Population in Kolkata, 2001.
Source: Based on data from Census of India 2001.

lighting. Date presented in Table 6.2 suggests that there is a significant spatial variation in terms of piped water supply within the houses. In Ward 58, only one-third of the households have piped water supply inside their houses while in ward 29 (Narkeldanga), about 70 per cent of the households have this facility. In Metiabruz cluster, Ward 137 (Metiabruz) is well endowed in terms of piped water supply as about 71 per cent of

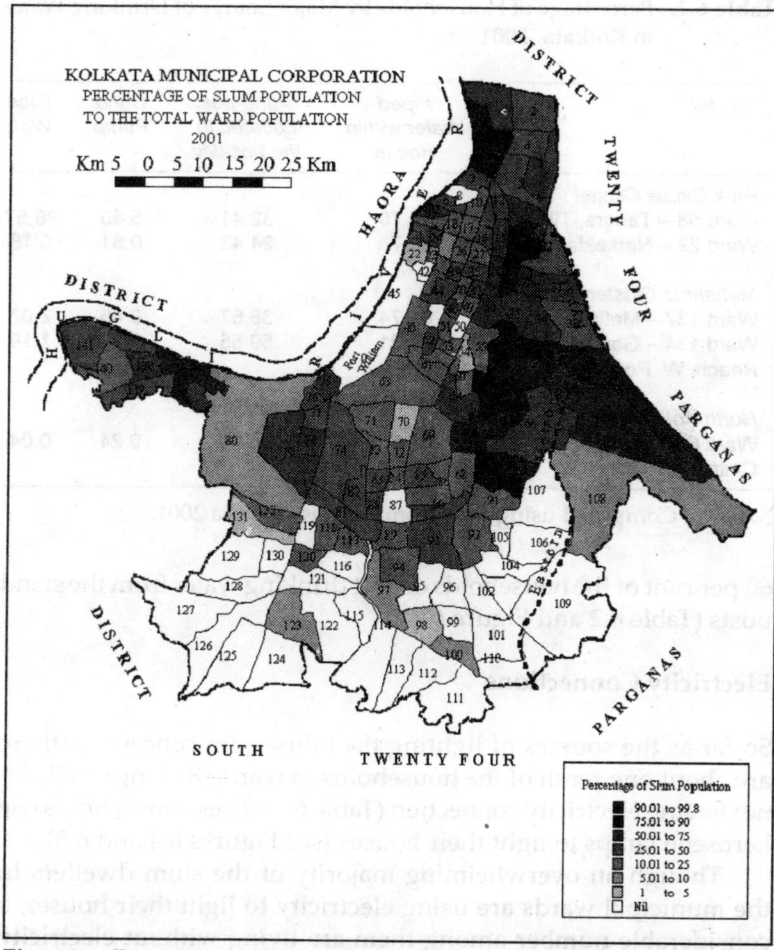

Fig. 6.2: Percentage of Slum Population in Wards of Kolkata Municipal Corporation, 2001.

Source: Based on data from Census of India 2001.

the households are having this facility but in Ward 134 (Garden Reach) only about one-third of the households have this facility. The piped water supply in Ward 6 (Cossipore) is far below the required levels as only 63.63 per cent of the households have such facility. However, in terms of piped water supply, Ward 134 (Garden Reach) has the most inadequate provisions as about

Table 6.2: Percentage of Households by Major Source of Drinking Water in Kolkata, 2001

Cluster	*Piped Water within House*	*Stand-post Located in the Locality*	*Hand Pump*	*Tube Well*
Park Circus Cluster				
Ward 58 – Tangra, Tiljala	34.70	32.41	5.40	26.57
Ward 29 – Narkeldanga	68.78	24.43	0.61	5.18
Metiabruz Cluster				
Ward 137 – Metiabruz	60.74	36.67	0.45	2.02
Ward 134 – Garden Reach, W. Port	37.21	59.55	0.77	1.19
North Kolkata Cluster				
Ward 6 – Cossipore, Chitpore	63.63	35.01	0.24	0.04

Source: Computed using data from Census of India 2001.

60 per cent of the households collect drinking water from the stand posts (Table 6.2 and Figure 6.3).

Electricity Connections

So far as the sources of lighting the houses are concerned, there are about one-tenth of the households in ward 58 (Tangra-Tiljala) not having electricity connection (Table 6.3). These households use kerosene lamps to light their houses (see Figures 6.4 and 6.5).

Though an overwhelming majority of the slum dwellers in the municipal wards are using electricity to light their houses, a considerable number among them are living without electricity connection. One-tenth of the households in Ward 58 (Tangra, Tiljala) is still using kerosene to light their houses (Figure 6.5). The darkest picture is that there are a number of households without having any means to light their houses.

Distribution of Assets

An asset is considered very valuable for slum dwellers as they can use them at a time of distress. A perusal of Table 6.4 suggests that not many in the slums own even a source of entertainment such as

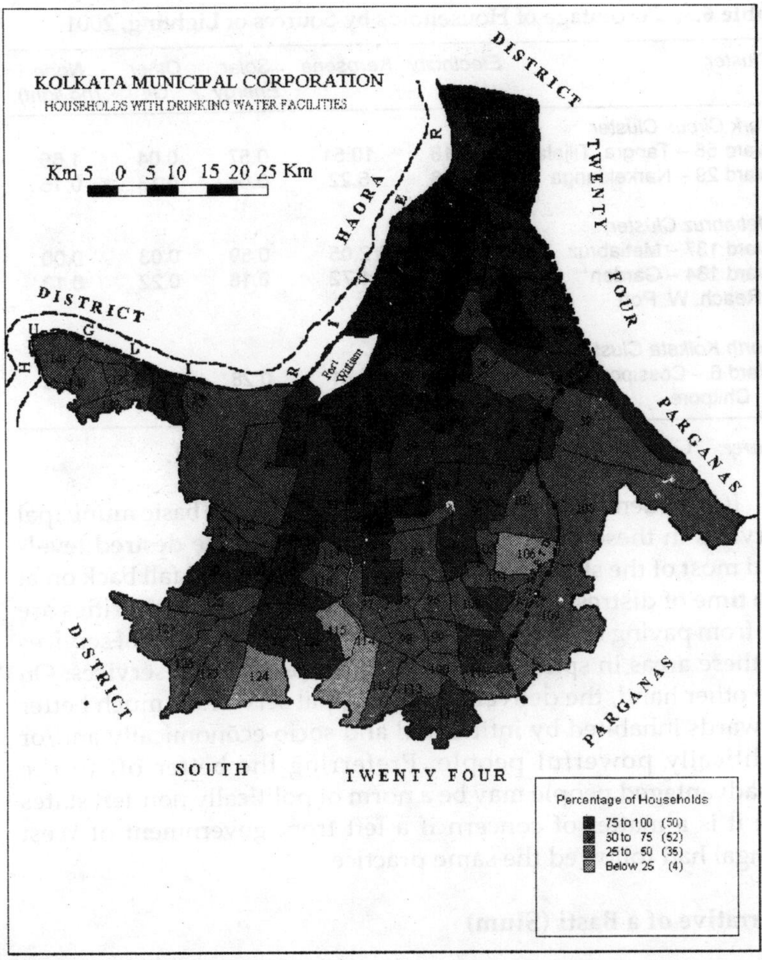

Fig. 6.3: Distribution of Households with Drinking Water Facilities.
Source: Based on data from Census of India 2001.

radio or TV. A telephone connection which is also considered to be a proof of residence, is nearly absent in most of the households. Depressingly, there are about one-third of the households in Ward 58 (Tangra, Tiljala), Ward 29 (Narkeldanga) and Ward 6 (Cossipore) not having any of these minimal assets such as radio, television, telephone, bicycle, motorbike, car/jeep/van.

Table 6.3: Percentage of Households by Sources of Lighting, 2001

Cluster	*Electricity*	*Kerosene*	*Solar Energy*	*Other Oil*	*None (no light)*
Park Circus Cluster					
Ward 58 – Tangra, Tiljala	87.18	10.51	0.57	0.04	1.65
Ward 29 – Narkeldanga	93.28	6.22	0.04	0.13	0.15
Metiabruz Cluster					
Ward 137 – Metiabruz	96.97	2.05	0.59	0.03	0.00
Ward 134 – Garden Reach, W. Port	97.66	1.72	0.18	0.22	0.13
North Kolkata Cluster					
Ward 6 – Cossipore, Chitpore	91.82	7.81	0.28	0.03	0.06

Source: Computed using data from Census of India 2001.

It is evident from the above discussion that the basic municipal services in these slum wards are much below the desired levels and most of the slum dwellers do not have assets to fall back on at the time of distress. It suggests that the municipal authorities are far from paying required attention in making provisions of services in these areas in spite of severe deficiencies of such services. On the other hand, the delivery of municipal services is much better in wards inhabited by influential and socio-economically and/or politically powerful people. Preferring the better off to the disadvantaged people may be a norm of politically non-left states but it is a matter of concern if a left front government of West Bengal had followed the same practice.

Narrative of a Basti (Slum)

The following narrative of a *basti* (slum), predominated by Muslims, gives an idea about the impact of changes that has taken place in the economic landscape of Kolkata and also a reflection of educational scenario.

The Priya Manna basti (PM basti) has been at the receiving end of the changes that have been taking place in the economic landscape of the city of Kolkata since the closure of the jute mills. The jute mills in and around Howrah used to provide employment to a large number of residents of this *basti* and the closure of several such mills have displaced a large population from their earlier

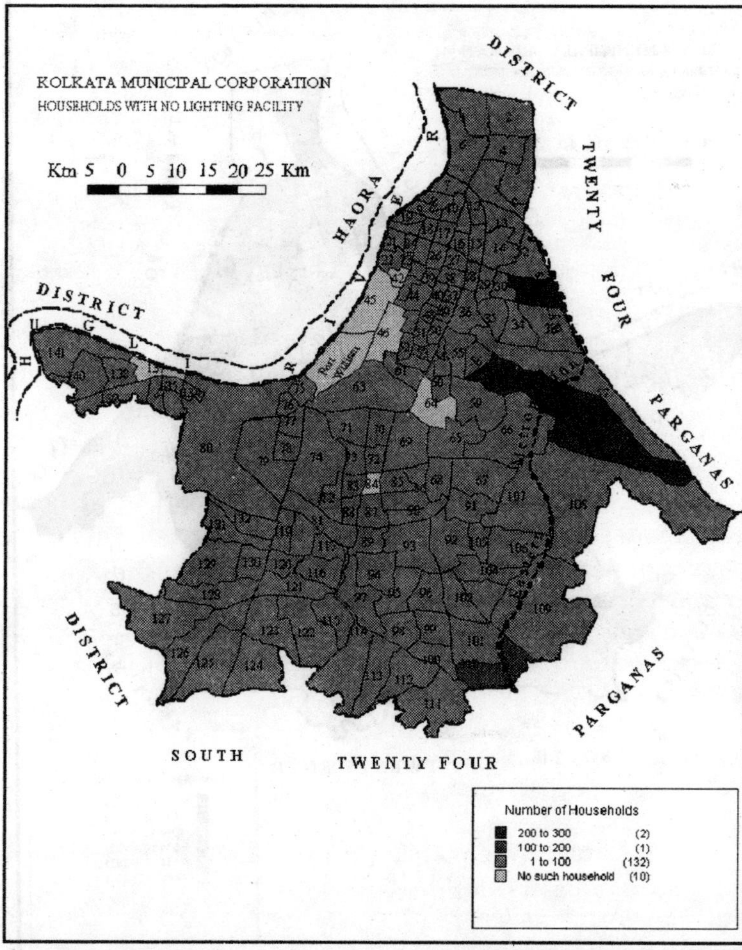

Fig. 6.4: Distribution of Households without Lighting Facilities, 2001.
Source: Based on data from Census of India 2001.

sources of livelihood. Since an alternative to their earlier occupation has been difficult to find, the economic condition of these residents has become more precarious. It has been as difficult for the residents to apply any strategy to cope with the stress arising from their loss of employment given their low levels of education and the lack of support from the government. Presently, a number of *basti* dwellers have managed to find low income, irregular jobs

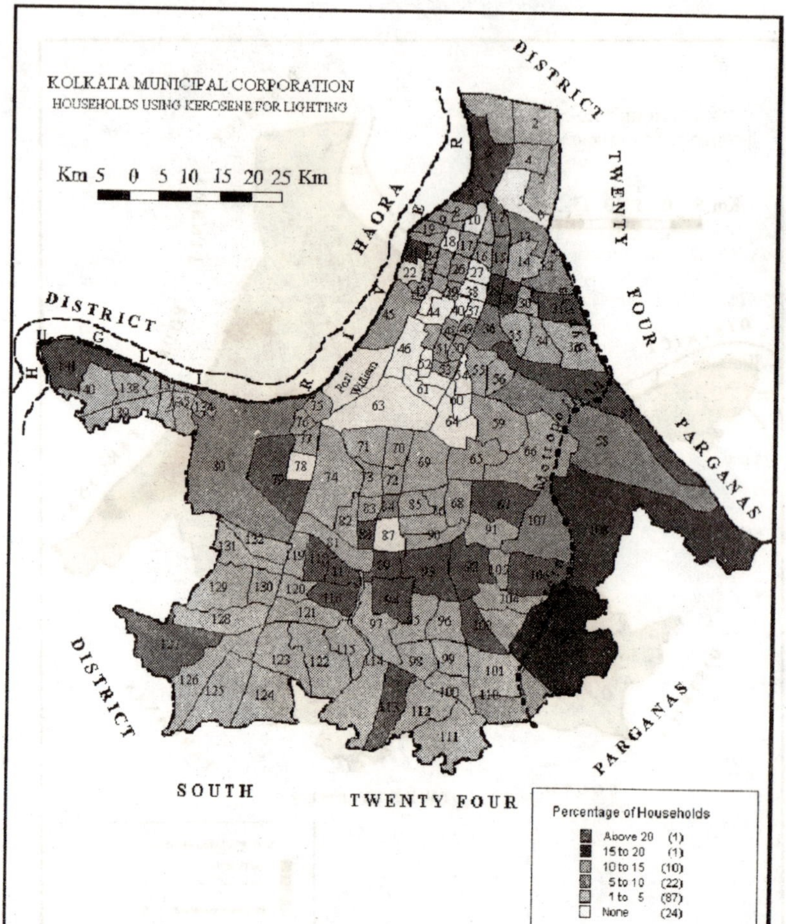

Fig. 6.5: Distribution of Households using Kerosene for Lighting in their Houses, 2001.

Source: Based on data from Census of India 2001.

in the informal sector or have become so-called 'self-employed' to sustain themselves with minimal levels of income. A large number of such self-employed people are engaged in lower end economic activities such as vegetable selling or owning a *paan dokan* (betel shop).

Table 6.4: Percentage Distribution of Households Owning Different Assets

Cluster	*Radio*	*TV*	*Telephone*	*Bicycle*	*Motorbike*	*Car etc.*	*None of these*
Park Circus Cluster							
Ward 58 – Tangra, Tiljala	42.11	49.24	10.71	31.53	6.85	2.81	30.15
Ward 29 – Narkeldanga	27.91	60.82	8.84	6.69	2.60	1.35	30.78
Metiabruz Cluster							
Ward 137 – Metiabruz	38.16	58.57	9.13	23.56	3.45	2.20	27.72
Ward 134 – Garden Reach, W. Port	40.45	66.03	7.97	26.63	4.57	1.87	21.96
North Kolkata Cluster							
Ward 6 – Cossipore, Chitpore	43.12	55.13	10.38	12.66	3.11	2.52	31.10

Source: Computed by Author from Census of India 2001.

In this context, a household survey was conducted in the *basti* during the last quarter of 2008. The survey covered 100 households with a population of 758. Since PM Basti is very big and for the sample to be representative, the *Basti* was divided into four spatial clusters. 25 households from each of these clusters were surveyed. It was also tried to ensure that the households located on the different lanes within a particular cluster were fairly represented. An interview schedule was developed and administered for the household survey. The interview schedule had both closed and open ended questions. The research assistants for conducting the survey were recruited from the slum itself. All of them had earlier experience of working with the community and they also had prior experience of conducting a household survey.

Demographic Characteristics of the Surveyed People

Total 758 people living in 100 households were surveyed, out of which 353 were women and 405 men. The *basti* has an unfavourable sex ratio (number of females per 1000 males is 872). The low sex ratio of the *basti* may be explained by the male selective immigration of the people, coming from the rural areas. A large number of *basti* dwellers live in single room units. Sometimes 10 people may be residing in a single room. Consequently, the household size is very large, that is 7.58.

About 4.0 per cent of the population belonged to the age group of below 4 years. The school going age group (5-19) children constituted about 45 per cent of the total population. A large majority of the population belonged to the working age group of 15-59 years: 63.2 per cent of the population belong to this age group. Only 5 per cent of the population belong to the older age-group of above 60 years.

Educational Scenario

The literacy rate in the *Basti* is about 70 per cent which is impressive considering lower literacy rates in the low income areas. However, the disparity in the male-female literacy rate is significant. The female literacy rate is 66.10 per cent as against 72.20 males.

Though the literacy rate is appreciable, the educational levels of the literate people are very low. About one-third of the literates are educated up to only primary level (4th standard) while another 45 per cent is educated up to middle level (7th standard). That means about three-fourth of the literates could not go beyond 7th standard. Those who could complete secondary level of education (10th standard) constitute about one-tenth of total literates. A mere 5 per cent has completed graduation (Table 6.5).

Table 6.5: Educational Levels of the Literate Population

Educational Levels	*Total*	*Male*	*Female*
Primary	148 (31.4%)	86 (32.2%)	62 (30.2%)
Middle	215 (45.6%)	115 (43.1%)	100 (48.8%)
Secondary	52 (11.0%)	31 (11.6%)	21 (10.2%)
Senior Secondary	31 (6.6%)	14 (5.2%)	17 (8.3%)
Graduate	24 (5.1%)	19 (7.1%)	5 (2.4%)
Ph.D.	2 (0.4%)	2 (0.7%)	–
Total	472 (100.0%)	267 (100.0%)	205 (100.0%)

Source: Household Survey.

Educational Deprivation of the Children

The number of children in the sample population of *basti* in the age-group of 5 to 19 years is 330, out of which 172 are boys and 158 girls. Though these children belong to the school going age group, a large number of them are not attending any school. About 15 per cent of the children in the age-group of 5-19 years have never entered the door of a school and about one-third though entered school, have dropped out. In other words, only about one-half of the children in this age group are continuing in school while the other half is out of school (see Table 6.6). It is worth mentioning that the percentage of girls, who have never gone to school, is lower than that of boys and conversely, the percentage of girls, who are continuing in school, is higher than that of boys. This gender differential may be explained by two separate reasons for boys and girls. Firstly, the boys bear a higher burden of working at an earlier age than the girls and secondly, the marriage prospects of educated girls are considered to be better than those who are not educated.

Table 6.6: Schooling Status of Children in the Age Group of 5-19

Current status	*Number*			*Percentage*		
	Total	*Boys*	*Girls*	*Total*	*Boys*	*Girls*
Never went to school	49	29	20	14.85	16.86	12.66
Continuing in school	177	87	90	53.63	50.58	56.96
Went to school but dropped out	104	56	48	31.52	32.56	30.38
Total	330	172	158	100.00	100.00	100.00

Source: Household Survey.

An analysis of the data on dropped out children at different stages of education suggests that the highest percentage of children abstain from going to school at middle level followed by primary level. In fact, the percentage of drop outs at the middle level is double that at primary level, that is 53.85 and 27.88 per cent respectively. It suggests that most of the drop outs occur either at middle or primary level. A little below one-fifth of the drop outs occurs at the high school level. An investigation of drop outs by their gender throws some interesting findings. The percentage of girls who dropped out at Primary level is about half as that of boys (See Table 6.7). On the other hand, the percentage of girls who dropped out at high school level is double as that of boys. It suggests that a number of girls are able to continue their education up to High School which may not be true for the boys.

Table 6.7: Stages of Schooling at which Children Dropped Out

	Number			*Percentage*		
	Total	*Boys*	*Girls*	*Total*	*Boys*	*Girls*
Primary (Classes I-IV)	29	21	8	27.88	37.50	16.67
Middle (V-VIII)	56	28	28	53.85	50.00	58.33
High School (IX-X)	19	7	12	18.27	12.50	25.00
Total	104	56	48	100.00	100.00	100.00

Source: Household Survey.

An investigation into the age-specific enrolment ratios suggests some interesting findings. Although both the boys and girls have a similar enrolment ratio of about 81 per cent at an early age cohort of 5-9 years, in the following age group of 10-14 years, the enrolment ratio falls down sharply to 67.11 per cent for the boys while it improves to 86.36 per cent for the girls. The enrolment ratio for both boys and girls takes a huge plunge in the next age cohort of 15-19 years (see Figure 6.6). The enrolment ratio remains a mere 14.29 per cent for the boys while in the case of girls it is reported to be 19.70 per cent. The age-group of 10-14 years appears to be a watershed in terms of continuation of education. Most of the school going children are unable to continue their education beyond the age of 14 years.

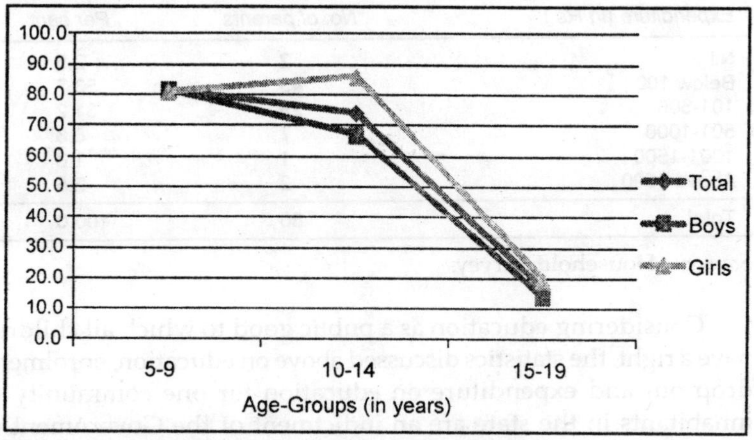

Fig. 6.6: Age-Specific Enrolment Ratios in Kolkata.

Source: Household Survey.

Cost of Education Borne by Parents

An overwhelming majority of the parents in the *basti* bear the cost of educating their children, incurring from Rs. 11.67 per child per month (pcpm) to Rs. 1,666 pcpm, towards education related costs for their children (Table 6.8). However, the majority of parents are concentrated towards the lower end of this expenditure spectrum. A perusal of Table 6.8 shows that about half the parents are spending less than Rs. 100 per child every month and one-third are spending between Rs. 101 and Rs. 500. As opposed to such low expenditures made by the vast majority of parents, there are a few who can afford to spend a substantial amount for their children's education. In fact, there are two parents who are spending more than Rs. 1,500 per month for their children's education. Based on this analysis, we may say that even those in the low income bracket of the *basti* have to spend some amount from their meagre incomes to support the education of their children. Thus, we may raise the question whether primary education is really free as claimed by government or parents have to spent money for one reason or the other to educate their children.

Table 6.8: Monthly per Child Expenditure made by Parents on Education

Expenditure (in Rs.)	*No. of parents*	*Per cent*
Nil	2	2.5
Below 100	43	53.7
101-500	25	31.2
501-1000	7	8.8
1001-1500	1	1.3
Above 1500	2	2.5
Total	80	100.0

Source: Household Survey.

Considering education as a public good to which all children have a right, the statistics discussed above on education, enrolment, drop out and expenditure on education for one community of inhabitants in the state are an indictment of the Government of West Bengal's failure to ensure education to the children. This is a matter that demands urgent attention and redressal both by the government and civil society.

Conclusion

Most of the slum dwellers are concentrated in certain pockets of the city and they have very limited access to public goods and services and that has led to various kinds of social and spatial exclusions. Since a majority of Muslims live in slums, they have been experiencing multiple disadvantages such as poor living conditions, stigmatisation due to their place of living, inadequate and inferior educational facilities and displacement from their traditional occupations. All these disadvantages individually or together are contributing towards declining socio-economic conditions of Muslims in the city and excluding them from the mainstream development. However, the existing spatial segregation in the city may be a combined outcome of historical process of in-migration and the institutional policies followed by colonial government as well as the successive West Bengal governments in the post-independent India. During the colonial period, the British very consciously organised the space and created segregation based 'master-client' and/or white-brown population. Thus, there were areas within the city marked exclusively for white/British population while the next horizontal layer was occupied by those serving them (the brown population). The departure of the British vacated the space for the Bengali *Bhadroloks*. As the post-colonial period witnessed the development of the city owing to industrialisation, a large number of unskilled, illiterate or semi-literate rural migrants from the hinterland flocked to the city in search of employment and they ended up living in slums without proper living conditions.

The educational facilities located in the Muslim concentrated areas, and availed by the majority of their students, suffer from inadequate infrastructure and teachers that leads to lower educational standards of the students. As a consequence of this, it becomes very difficult for most of them to seek admission in institutions offering higher education and ultimately they end up without proper education and/or lower job prospects. The discussions with a number of Muslims reveal that the government officials managing the educational affairs are very apathetic to intervene and support in raising the standards of such schools, firstly, because these are Urdu medium schools and secondly, because most of the students and teachers are Muslims.

REFERENCES

Appasamy, Paul, S. Guhan, R. Hema, Manabi Majumdar and A. Vaidyanathan. 1995. "Social exclusion in Respect of Basic Needs in India", in Rodgers, Gerry, Charles Gore and Jose B. Figueiredo (eds.). *Social Exclusion: Rhetoric, Reality, Response*, pp. 237-252. Geneva: International Institute of Labour Studies.

De Haan, Arjan. 1994. *Bibliographical Review on Social Exclusion in South Asia*. Geneva: International Institute of Labour Studies.

Desai, Anjana P. and Aparajita De. 2002. "Communal Confrontations and Changing Social Space in Indian Cities: Case of Ahmedabad", *Ekistics*, Special issue on Architecture & Features II.

Firdos, Sohel. 2008. "Mapping the Disadvantaged Population in Kolkata City", Project on Socio-spatial exclusion in Kolkata, Centre for Studies in Social Sciences, Patuli, Kolkata.

Firdos, Sohel. 2009. "The Politics of Urban Service Provisions in the City of Kolkata", Paper presented at International Workshop on The Voice of City Dwellers in Urban Governance, Participation, Mobilisation and Local Democracy, organised by Centre de Sciences Humaines (CSH), January 12-14, Mumbai.

Mahadevia, Darshini. 2007. "A city with many borders: beyond ghettoisation in Ahmedabad", in Annapurna Shaw (ed.), *Indian Cities in Transition*, pp. 341-389. Hyderabad: Orient Longman.

Mohan, John. 2002. "Geographies of Welfare and Social Exclusion: Dimensions, Consequences and Methods". *Progress in Human Geography*, Vol. 26, no. 1, pp. 65-75.

Nayak, Pulin. 1994. *Economic Development and Social Exclusion in India*. Geneva: International Institute of Labour Studies.

Rodgers, Gerry, Charles Gore and Jose B. Figueiredo. eds. 1995. *Social Exclusion: Rhetoric, Reality, Response*. Geneva: International Institute of Labour Studies.

Sen, Amartya. 2000. *Social Exclusion: Concept, Application, and Scrutiny*. Social Development Papers No. 1, Asian Development Bank, Manila.

Sen, Amartya. 1981. *Poverty and Famines: An Essay on Entitlement and Deprivation*, Oxford: Clarendon Press.

7

Patriarchy, Purdah and Accumulation: A Study of Chikan Industry and Muslim Workers of Lucknow

Paaritosh Nath

Handicrafts in India are neither just a means of livelihood, nor can they be solely perceived as an art form—a mere manifestation of the artists' creative impulses. They comprise both these aspects and much more. For the artisan, handicrafts are as much about economics as they are about aesthetics. They are a part of his/her roots; a symbolic representation of childhood folk tales, regional customs and songs long lost from memory. Representing generations of familial skills and secrets; they are at the same time a crucial source of income and a distinctive marker of identity. Handicrafts moreover are not only about the artisans. They tell us, the non-artisans, about our own lives, our habits, our Gods and our mores. They have remained an integral part of our social and cultural matrix from the earliest of times and continue to do even today. Undoubtedly, social and cultural systems have proven to be the driving force behind the persistence of Indian handicrafts in the twenty-first century. Ironically enough, they also serve as the most potent agents of accumulation within the handicraft industry. After all, handicrafts are indeed as much about economics as they are about aesthetics!

Forming a part of the informal sector of our economy, the handicraft industry today is no longer considered to be outside the ambit of capitalist accumulation or market exchange. In fact, it has become an intrinsic part of both (Harriss-White, 2003).

Littlefield (1978) in her paper on the hammock industry in Yucatan, Mexico, provides a vivid account as to how societies with conventional or non-capitalist mode of production are increasingly being incorporated into the global capitalist system. Merchants and 'entrepreneurs', driven by the lure of high profits, have 'proletarianised' the once independent artisans and 'now control both the means needed by peasants or artisans to undertake production and the conditions for the reproduction of production' (Littlefield, 1978: 495). Third World artisans, increasingly producing for the global market, are more than ever exposed to the daily vagaries that such markets have to offer (Scrase, 2003). Accumulation in the handicraft sector necessitates the continued control and exploitation of the artisans (for their surplus value) by the merchant classes. In order to come to terms with the nature of exploitation of the artisans in the handicraft economy, it is crucial to understand how these accumulating classes are able to exploit others and to induce other classes to accept as in their interests the requirements for continued capital accumulation (Leys, 1996).

This chapter seeks to unravel these very processes of accumulation and demonstrate the role cultural and social institutions play in creating and maintaining an ideological/psychological base on which 'an entire phase of exploitation can be built' (Mies, 1981). Specifically, it will unravel how capitalist accumulation is brought about by the overexploitation of female artisans using: (a) the cultural connotations of gender and gender roles in relation to labour (what Mies calls the 'sexual division of labour') and (b) cultural and ideological mechanisms such as the '*purdah*' that call for the domestication of female artisans, in the process, ensuring a cheap and available supply of home based workers. This chapter examines these formulations in the light of the conditions of life and work of women Chikankari workers of Lucknow, Uttar Pradesh.

The issue, then, concerning the major portion of this work is the role that culture has to play in the processes of accumulation of capital. To what degree can norms and prejudices determine the economic fortunes of individuals? How are social institutions such as gender manipulated by capitalists to exploit surplus labour value?

Most of the available literature on caste, religion and gender in India as Harriss-White (2003) correctly points out, typically

defines the above as components of civil society—often lumped together as 'networks' or 'social capital'. Gordon (1982), Kotz (1994) and other members of the Social Structures of Accumulation (SSA) School, however, argue that these institutions need to be seen as separate structures of accumulation, each working with an in-built logic of their own. Social institutions such as race, gender and culture, in fact, have an immensely important role to play in capitalist accumulation. Norms and attitudes attached to such institutions, they reckon, are the means by which the division of labour, segmentation and surplus distribution is legitimized in the socio-economic complex within which they operate.

The SSA school critiques mainstream attitude towards accumulation for its reductionism and ineptness insofar as it tends to restrict capitalism to the market which, in turn, is further restricted to the 'pure economic domain' determined simply by the intersections of supply and demand (Harriss-White, 2003). In doing so, it ignores the crucial part that social institutions and the norms and practices attached to them play in determining the supply, the costs and regulation of the different factors of production. On the same lines, this chapter *via* a case study of the Chikankari workers, argues that processes of accumulation in this industry do not only take place solely in the market/workplace and in fact are more often than not determined by 'non- economic' factors/institutions that operate beyond it. As we shall observe in the following section, discrimination on the grounds of social and religious mores often entwine with gender prejudices to produce forms and types of exclusion unique to women artisanal groups.

The Lucknow Chikankari Industry

"Chikankari" is white floral embroidery performed (mostly) on white fabric; an art form that developed during the Mughal period in Dacca (present Dhaka), Calcutta (present Kolkata) and Lucknow. The word "Chikan" can trace its etymology to a Persian word derived from the Chic, which referred to the 'Jali' work carved on marble and wood. As one version of the story goes, the introduction of Chikankari to India can be credited to Queen Nur Jahan, the wife of Mughal emperor Jahangir who brought the artwork from Persia to India. Though there is no means of verifying whether there is any truth to this particular claim (one that is nonetheless popular amongst the artisan communities of Lucknow, the Persian influence to the art cannot be ignored. It was first made as a cheaper version of brocaded "jamdani" cloth, incorporating elements of eighteenth century European embroidery. Chikan those days was worked on such fine muslin cloth that such early examples have not survived today.

Chikan survives today only in Lucknow, produced mostly by Muslim women but controlled at the top by Hindu traders. Embroidered garments are made in stages, starting with fabric cutting and tailoring, followed by block printing and embroidering, and, finally, laundering. In all stages but one, the male specialists predominate. However, the embroidery stage is completely dominated by women. A report compiled by the Textiles Committee, Government of India titled, 'Art and Artifacts of Lucknow Chikankari' (2007:25) estimates that about "2.5 lakh artisans of seven different districts are earning their livelihood by practicing this craft, making Chikan embroidery in turn, the biggest cluster in the country". It is estimated that the artisans of Chikan embroidery are scattered in the villages of Lucknow district and other nearby districts in a radius of about 125 km. The districts covered by the craft are Barabanki, Unnao, Lakhimpur, Hardoi and Sitapur. Other districts where one can find artisans of Chikankari are Rai Bareilly, Shahjahanpur, Sultanpur and Faizabad (Report of Textile Committee, 2007).

In recent years, with the opening up of the export markets and the increase in the domestic market size that the industry now caters to, Chikankari has seen a phenomenal increase in its overall turnover. Though most *Mahajans* (merchants/traders) are extremely vary of announcing annual turnover that they make per annum, a good idea can, nevertheless, be made by taking into account the figures as put forth in a report on the same by the Ministry of Small Scale Industries, Government of India, New Delhi, in the year 2006-07.

As per the reports, the total turnover of the Chikan cluster was found to be around Rs. 1,000 crore per annum. Out of the five thousand or so enterprises engaged in the Chikan industry, five of them generated revenue between two to five crore a year; more than three hundred enterprises between rupees fifty lakhs to one crore. However, despite the increasingly growing demand for and popularity of "Chikankari", both within and outside Lucknow,

Table 7.1: Turnover of the Enterprises in the Chikan Cluster (2006-07)

Turnover (Rs. per annum)	*Number of Enterprises*
2-5 crores	5
1-3 crores	10
50-100 lakhs	300
20-50 lakhs	750
5-20 lakhs	2000
3-5 lakhs	2000
Total	5065

Source: SSI, Diagnostic Study of the Chikan Embroidery Cluster, Lucknow, Uttar Pradesh.

the embroiderers of this fine craft continue to live a life of poverty and destitution, compounded by their backwardness and seclusion. The Chikankari artisans of Lucknow remain one of the most poorly paid, vulnerable and abused artisans in the entire country. At times, they are forced to remain 'content' with less than three to four hundred rupees per month for a month's hard labour. Considering the fact that the State's minimum wage for unskilled work alone is rupees one hundred and fifty per day, it leaves little to comprehend the magnitude of their exploitation.

What accounts for such precarious existences of the embroiderers of this traditional art? The answers to this have to be sought in the organisational pyramid that the industry subscribes to and the relations of socio-economic exchange carried out between different actors in the production chain.

Production Value Chain of the Chikankari Industry

Located at the top most level of the organizational pyramid of the Chikan Industry are the entrepreneurs (Mahajans). These entrepreneurs have an exceptionally strong hold over the entire production chain of the industry. What is to be produced and in what quantities—the decision solely lies with these entrepreneurs who sub-contract and distribute the majority of the work to a number of value chain professionals. Most of these manufacturers belong to high caste Hindu families that have been traditionally engaged in the profession. These entrepreneurs perform the dual role of maintaining a liaison among the production chain and selling the final goods in the market. Apart from this category of entrepreneurs there also exist a number of retail shop owners who buy the finished product from the traders (who in this case act as wholesalers) in bulk and put the goods up for sale in their showrooms. Both the above categories of entrepreneurs are located at the end of the supply chain, with the former also finding themselves at the final stages of the production chain.

Next in the organisational pyramid of the Chikan industry, we have the middlemen or the *'bichoulias'*. The middlemen or agents, in the words of a senior government official of the Uttar Pradesh Export Promotion Bureau, constitute the 'backbone of the industry'. These are the *'bichoulias'* that receive raw materials as well as instructions from the entrepreneur and distribute work to various workers employed in the production process. Thus, the *bichoulias* act as links between the manufacturers and the sub-contracted washermen, printers, tailors etc. More importantly, the middlemen serve as the sole point of contact between the entrepreneurs and the embroiderers or the Chikankars. The *bichoulias*, it can be said, represent in more ways than one the informal and unregulated nature of this particular industry. These are the *bichoulias* that fix the rates of the products of chikankars and deliver the embroidered product to the sellers/bigger contractors. The wage rates of the workers are also decided by the *bichoulias*. The difference between the two (sets of rates) is pocketed by the agents and becomes their source of income.

It is the middleman's prerogative to choose the workers that will work for him/her. The workers employed are generally known to the agent having worked for the latter on similar consignments in the past. Depending upon the size and nature of the order, the workers and agents fix a date in the future when the embroidered work shall be ready for collection. The contract fixed is always *via* word of mouth and never written. In fact, in the entire production line, most contracts and sub-contracts established are oral in nature carrying no legal bindings upon them whatsoever. The oral nature of these contracts further puts the *bichoulia* in a position of power over the artisan who has to solely rely upon the former's assurances of fair and timely payment. Traditionally *bichoulias* used to be men, engaged in the task of running around, distributing work from house to house and collecting final products—activities that women in *'purdah'* were strictly forbidden to execute. However, in the last twenty years or so, the Lucknow Chikan industry has seen the emergence of a number of women as *bichoulias*. Most of them are former embroiderers who have started sub-contracting work themselves. They generally operate within the city, distributing work obtained from entrepreneurs to women who live near their neighbourhoods.

Below the *bichoulia* in the organisational hierarchy are the artisans or the chikankars. The artisan is the pivotal, indispensible figure of the industry; the craftsperson without whose knowledge and skill the industry shall simply cease to exist. It is the artisan's gift (of embroidery) that confers Chikankari its unique status and creates its own demand within the market. Surprisingly it is then to find them so lower down the order in the power structure of the industry.

The Lucknow Chikankari industry, as one can make out from the above, is a surviving example of the putting out system of production and herein, the author reckons, lays the crux of the problem. The 'putting out', according to Baishya:

> "Is a system of manufacturing where the raw materials are delivered out to so-called independent craftsmen who while processing the materials at different stages with their own instruments of production return the products to the raw material supplying agents" (1997: 51).

Mainstream literature defines the above as a system of domestic manufacturing that was prevalent in rural areas of Western Europe during the seventeenth and eighteenth centuries. Weber (1981) saw it as a phase symbolizing the transition from traditional handicraft production to modern manufacturing. Marx (1977) saw in the putting out system what he termed as the 'formal subordination of labour to capital'. Unlike the real subordination (of labour to capital) in this production process, wage earners still had considerable control over their own labour process. In most cases, the putting-out workers owned their tools, determined the length and intensity of their workday, and had few or no employees.

As Lazerson (1995) points out, as a precursor to the industrial revolution, the putting-out was expected to disappear once centralized manufacturing processes matured. Marx saw the putting out as a transitory form which would eventually be replaced by the factory system of production.

> "The cheapening of the labour power by sheer abuse of the labour of women and children, by sheer robbery of every normal condition requisite for working and living, and by the sheer brutality of overwork and night work meets at last with natural obstacles that cannot be overstepped.... So soon as this point is at last reached — and it takes many years — the hour has struck for the introduction of machinery and for the henceforth rapid conversion of the scattered domestic industries and also manufactures into factories" (Marx, 1977: 466).

Liberal economists still view the putting out system as suffering from poor workmanship, pervasive threat and spasmodic co-ordination; an economically inefficient system whose demise they assure shall come at the hands of the modern factory (Landes, 1969; Bythell, 1978). The logic of capitalism, institutional economists argue, necessitates the abandonment of the putting out as it fails to extract maximum surplus from the worker (Marglin, 1974).

The Chikan industry of Lucknow, however, right from its inception has operated on the putting out system of production and continues to do so even today. Despite the overall advent of technology in the textiles sector in terms of control over both the quality and quantity of the finished goods, not once has the existent production chain been threatened in any manner. In other words, the supposedly "transitional phase" over here is a lasting one. To be fair, the demise of the putting out system as predicted by Marx was laid out for the nations that were to undergo a complete sectoral transition from agriculture to industrialization—something that definitely has not occurred in India. In any case, what he did not foresee was the continued persistence of such production systems in other parts of the world, along with the factory system, and their expansion with the advent of the global capitalist system (Mies, 1981).

The question then to be raised is, what exactly are the contributing factors that ensure the persistence of the putting out system in the Chikankari industry? How do we come to terms with the fact that a supposedly inefficient system of production, in fact, is expanding and getting strengthened over the years? The presence of this sub-contracting system, can, to a certain extent, be ascribed to the availability of a large supply of workers (here artisans) who insist to work from home despite the fact that home based work clearly acts as a wage depressor. Though a number of factors such as responsibilities of the household, child care etc. contribute towards the continued supply of cheap home based labour, the author would like to lay emphasis on certain social/ cultural norms and customs deeply rooted in the patriarchal modes of production that contribute towards this, the most significant undoubtedly being the *'purdah'* system.

It is to be noted that the sense in which we use the term *'purdah'* (while defining its role in both accumulation and labour market segmentation) goes beyond its simplistic rather superficial association with the practice of 'hiding one's face and form' (Abdullah and Zeidenstein as quoted in Cain *et. al*, 1979).[1] According to Amin (1997), it is more useful to approach the *'purdah'* as the broader set of norms and customs that attempt to set standards of female morality. These norms have, in turn, far-reaching consequences as far as women's socio-economic exchanges are concerned, especially with men outside her household. Any violation of these fixed standards results in 'shame' and 'loss of honour' not only of the woman concerned but her family as well. Maintenance of the norms of the *'purdah'* more often than not necessitates their exclusion from the public sphere and provides for specific gender identities to labour. As Haque (2010) points out, the institution known as *'purdah'* is not merely a physical description of the veil or the homestead. He opines that the system has two forms—the visible that manifests itself in the form of 'seclusion, dress and segregation of the sexes' and the invisible that is ubiquitous in the underlying attitudes of society—the use of culture and tradition as instruments of control over women.

Most of the female embroiderers in the Chikan industry hail from low income Muslim households of Lucknow and its adjoining regions where the *'purdah'* system is in vogue even today[2]. Thus, a number of Chikan embroiderers live a highly restricted public life and are either not willing to or not allowed to conduct transactions in the market face to face. Because of limitations on their mobility and visibility in the public realm, they learn no other skills, and cannot pursue jobs in offices or factories. For them, the household based system of production proves to be the only possible means by which some income may be earned without compromising upon the codes of conduct laid out for them by society. Weiss (1992), in her study on the socio-economic condition of working women in Lahore states that the *'purdah'* system has a detrimental effect on a woman worker's self-image, the kinds of jobs she could do, and her bargaining position in economic relations. Much the same holds true for Chikan embroiderers. The system ensures that women embroiderers never enter the labour market and are forced to operate *via* the middleman. Lack of access to both, product and labour market, guarantees that they have incomplete, at times, incorrect information regarding the same. Villagers who do not even enter the urban markets as consumers of chikan products are the ones least informed about the market in Chikan. Thus, they have no clear-cut idea regarding the actual selling prices of the products that they embroider. More importantly, they are left clueless about the wages other embroiderers are receiving for the same amount of work. Both the situations described above are exploited by those higher in the organisational chain (whether it be the manufacturers, the agent or the both) in compressing the wages that should ideally accrue to the artisan. Physical segregation moreover, ensures that they cannot form labour unions/associations and apply pressure upon their employers to improve wage and working conditions.

Box 7.1: The *Purdah*-Poverty Nexus

Nisar Farooqui, (name changed) aged fifty years, hails from the old city area of Lucknow. Widowed twelve years back, she has been performing stitching and jali work ever since, in order to support her family of four. Like hundreds of other Chikankari workers, Nisar lives in a dingy one room apartment in Bhadevan Colony, just two kilometres away from the chikan centre of the city, i.e. Chowk. Like them, she too works from her own place and receives her orders from *bichoulias* working for different Mahajans located in Chowk. For the lengthy hours that she puts in, she receives on an average a compensation sum of seven hundred rupees per month. Nisar, we can say, in many ways is the average chikan worker that we come across in Lucknow save for the fact that she has a postgraduation degree in Sociology from the University of Lucknow!

Box 7.1 (Contd.)

A story that at once speaks of the degree of exploitation in the industry as well as the strictly segmented nature of the labour market (where even well educated women find it difficult to move out of pre-defined occupations), Nisar's tale of predicaments began as soon as her husband passed away leaving her a widow in her late forties to take care of three children. Initially, she tried moving out of her house and seeking employment elsewhere on the basis of the degree she had acquired before marriage, but employment opportunities came far and few. "No one", she recollects, "wanted to hire an old, Muslim widow to work in their firms or offices. Not only was I handicapped because of my sex but also because of my age and religion. The only opportunities that I received required door to door salesmanship and I could not travel that much in this age." There was resistance from the community as well. As she recalls, "my relatives and friends advised me that it did not look proper for a woman of my age and background to work outside like a common labourer."

Unable to find a decent, fixed source of employment, she began practicing chikan work learning the skill at a centre opened up by one of the *bichoulias* in her locality. Chikankari was the only profession where she could fend for her family and still maintain the dictates of the *'purdah'* which her relatives wanted. Initially starting off with coarse, *'mota kaam'* she has slowly broadened her knowledge of the stitches involved. Today, she can perform most of the major stitches. Yet, her monthly earnings remain extremely low.

She mentions somewhat bitterly, "I know that I am getting lower wages. In fact, most other chikankars I know receive higher remuneration. But the fact is, I am totally dependent upon this income and the *bichoulias* know that." When asked if she would not mind becoming a *Bichoulia* herself, she replies, "But that requires some amount of initial capital which I cannot afford."

She says that she has tried again and again to leave this profession and get what she terms 'a proper job' but all efforts

(Contd.)

Box 7.1 (Contd.)

have been in vain. Her university degree, she laments is useless. The family manages from whatever she and her oldest son (who has now started working) are able to bring back home. Today, the family's total monthly income stands at three thousand rupees a month.

Trapped in a profession that is for her a necessary yet undesirable line of work, she seems absolutely prepared to leave the industry if any better opportunities happen to come her way. She is even more certain that she shall not allow her only daughter to move into chikankari. The latter who is still studying in college, "should not", reckons the mother, "see the days I've seen. Rather she should utilize her education into building a substantive stable future for herself."

Nisar's story demonstrates how cultural and societal norms often force women to take up work far below what they are actually qualified for. In her case, it was the persistent pressure from her relatives and other social groups to remain at home and maintain the dignity associated with the *'purdah'* rather than move outside and search for work that brought her family to mere penury levels. Her tale, moreover, speaks of the present labour market conditions where women are still predominantly 'wanted' in certain professions and where access to other professions is denied by the patriarchal code. Most of the professions open for women, moreover are those which either allow for greater capital accumulation (by siphoning off more and more surplus labour) such as home based textile industries or those that directly participate in the upkeep and maintenance of labour supply. The Chikan industry of Lucknow is no different. The entire production system is structured in such a manner that it absorbs even highly qualified women like Nisar and makes them work for mere token amounts. Heavy surplus value extraction ensues.

Source: Based on interview by the Author.

Even for women workers not compelled to observe the *"purdah"*, the onerous responsibilities associated with managing large households limit their ventures outside the house. Yet the reason why this chapter chooses to specifically highlight the social norm of the *'purdah'* is for the following counts: (a) Many women who are not burdened with much household work or desire to enter into other professions are not allowed to move beyond the norms of 'decency' and 'honour' and take up work outside the house. The case of Nisar (see Box 7.1) was just one of the many that the author came across in the district of Lucknow. (b) *'Purdah'* as mentioned above ensures exclusion not only from the regular wage market but tends to restrict interactions of any nature. The women workers of Chikankari, as a result, have little information regarding the market conditions prevailing in the industry. This is especially pertinent in the rural areas adjoining the city where the agent at times is her only reference point with the outside world. Both these points attached to the social practice of the *'purdah'*, as we shall see below, serve as instruments for accumulation of surplus value in the Industry.

Specific social norms demanding domestication coalesce with ubiquitous gender discrimination to result in an exploitation that is doubly enforced. It is thus, not only the practice of the *"purdah"* but also the fact that the chikankari workers are women (irrespective of their social-cultural identity) that contributes towards their exploitation. A number of cross-cultural studies on the employment of women in the informal sector (of which household production system is a part) seem to suggest that cultural constructions of gender and gender relations often serve as important ideological tools to extract surplus value from women workers. The cultural roles associated with both men and women (the former linked with transcendence, the latter with domestic permanence) not only give way to unpaid domestic work but also have an effect on the kind of work and wages that female as paid labourers get.

Wilkinson-Weber (1997) highlights this in her analysis of the female embroiderers of the Lucknow Chikankari industry. The embroiderers are seen as 'classic home workers' who balance wage labour work with domestic chores such as cooking, cleaning, and looking after the children etc. The traditional roles assigned to them are that of the mother and wife. The same classification, made by their employers, casts them in the role of "leisure-time embroiderers", unworthy of the wages and attention given to "real" workers. Thus, is the justification for the extremely low incomes that they are given.

A significant third factor that contributes towards the precarious condition of the Chikankars has to be located in the general sense of apathy, discrimination and at times xenophobia that society harbours towards these women. As with Nisar's case, a number of Muslim chikankars when interviewed by the author remarked that employers (outside the Chikankari industry) were quite 'hesitant' to employ Muslim women in their service. Some of these women were even willing to defy the *'purdah'* but were still denied employment.

Exclusion from formal wage employment is a reality that Muslims across the country and not only Lucknow have to contend with. According to Kazi (1999), more than sixty per cent of Muslim women workers are self-employed, much higher than their Christian and Hindu counterparts. The Muslim share in public employment is, however, less than 3 per cent. The presence of such high self-employment rates coupled with low levels of participation of Muslim women as salaried workers indicate their marginal presence as workers in the formal economy. Even women like Nisar who are educationally qualified often face numerous barriers to gainful, 'formal' work due to the negative prejudices and stereotypes attached to the Muslim community that are in vogue 'in an increasingly xenophobic Indian society' (Dale *et. al*, 2002).

Going by the Sachar Committee Report (2006), Muslims, in general, are one of the most marginalised communities of Indian society. If we take the labour market for example, only 18 per cent of Muslim men are in regular employment as against the twenty-five per cent Hindus. The overall work participation rate of Muslim men (48%) is lower than even that of Dalit men (53%). Muslims are distressingly underrepresented in government jobs. Data compiled by the Sachar Committee (where the Muslims' population share is 13.4 per cent of the total population in the country) shows that they hold a tiny 5.7 per cent of government jobs. In States with a high Muslim population (Uttar Pradesh, Bihar and West Bengal), this ratio is less than a third of their population share. Undoubtedly, when compared to their non-Muslim counterparts, Muslim men lag far behind in most socio-economic indicators thus highlighting their overall impoverished position.

Yet, it is also recognised (Hasan, 1994) that within the community they enjoy a far better position than the women who in a patriarchal set up are generally controlled and dominated by them. It is to be noted that partly Muslim men are themselves responsible for the overall exclusion of Muslim women from various crucial facets of life especially those in the public sphere which they (Muslim men) themselves enjoy.

The above being said, it must be noted, however, that the nature of exclusion of Indian Muslim women cannot be explained only by taking into account their disadvantaged position as women in a patriarchal set-up. If this were the case, the nature of exclusion that Muslim women regularly face would have been similar to the ones being faced by the women of the other religious communities. This, however, is far from the case. A considerable amount of work on the same (Raju, 1993; Sethuraman, 1998; Das, 2005; Siddique, 2005) informs us, that Muslim women are at a clear disadvantage when it comes to performance on certain key socio-economic indicators when compared to their non-Muslim counterparts. Thus, it is not only on account of being 'Muslim' or 'Women' that they face such hardships; rather it is their status as 'Muslim Women' which accounts for their exclusion.

Income from Chikankari

A study by Mathur (1975) in the mid-seventies revealed that the average chikan worker earned close to twenty-eight rupees per month. Ahmed and Dixit (1991) in their report on the 'Role and Status of Chikankari Workers of Lucknow' calculated the income of the chikankar to be close to five rupees per day. This would roughly result in a monthly income of Rs. 100-150 per month depending upon the number of days employed. Another survey conducted by "Ekatra" in 1994 estimated that the average monthly income from chikankari varies between less than a hundred rupees per month to about six hundred rupees per month. Higher levels in the range given above were generally earned by chikankar-cum-*bichoulias* or by the chikankari masters. The Singh and Jafri Report (2008) calculated average monthly income of the worker as being Rs. 575.

In Table 7.2 we can see that the maximum respondents (approximately 52%) earn between Rs. 400 to Rs. 800 per month.

A considerable number earn between Rs. 800 to Rs. 1,000 as well. Moreover, it can be observed that variations in chikankari income by geographical locality (Urban/Rural) apply more to extreme income groups than the middle ones. Data showed that sixteen per cent of rural respondents earn between Rs. 200 to Rs. 400 as compared to four per cent in the urban areas. Similarly, eight per cent of the urban based chikankars earn more than Rs. 1,600 as compared to none in the rural areas. The present study estimates that the average monthly income of Chikankari workers is about Rs. 695 per month, which is approximately Rs. 23 per day (Table 7.3). With women burdened with the responsibility of sharing household expenditure costs, at times even having to manage the entire household on their shoulders, such a paltry sum is far from sufficient.

Looking at the wage rates being given to workers since the 1970s, it is clear that though in nominal terms there has been a rise in wages but in real terms it has pretty much remained stagnant. Workers are mostly paid on piece rate and the total amounts they receive are subject to a number of factors, not all being economic in nature.

Gender and Culture as 'Structures of Accumulation'[3]

As mentioned above, the low wages accruing to the chikankars can be pinned down to the cultural/social constructions of gender difference (in society) that are often used as ideological tools to extract surplus value from female labourers. The manufacturers and the agents see chikankari work as being done during the '*fursat*' (leisure) hours of the day when the women of the household are free from domestic responsibilities and can devote time to their orders. The Chikankar's work is still deemed part-time and unworthy of the wages and provisions given to 'proper' workers. The embroiderers, balancing out work with familial chores, are more often than not identified with the roles of a mother and wife; stereotypes that render them undeserving of compensation levels that ought to be paid to 'real' (read male) workers.

This impression of chikankari being a part-time activity performed by women whose primary job is to ensure the maintenance of their households has severe ramifications on the socio-economic condition of these workers. The manufacturers and

Table 7.2: Monthly Income of Chikankari Workers

Average Monthly Income (in Rs.)	*No. of Rural Respondents*	*Rural Respondent %*	*No. of Urban Respondents*	*Urban Respondent %*	*Total Respondent*	*Total Respondent %*
< 200	–	–	–	–	–	–
200-400	4	16	1	4	5	10
400-600	6	24	7	28	13	26
600-800	8	32	5	20	13	26
800-1000	3	12	5	20	8	16
1000-1200	3	12	3	12	6	12
1200-1400	1	4	1	4	2	4
1400-1600	–	–	1	4	1	2
>1600	–	–	2	8	2	4
Total	25	100	25	100	50	100

Source: Field Survey.

Table 7.3: Average Monthly Income of Chikankari Workers

Chikankars	*Lowest (in Rs.)*	*Highest (in Rs.)*	*Average (in Rs.)*
Rural	200	1200	592
Urban	250	2200	798
Total	225	1600	695

Source: Field Survey.

entrepreneurs when questioned about the low wages provided to the chikankars generally defend themselves on the grounds that the money provided to them is at par with the number of hours that have gone into the work. The majority of them claim that these workers work for no more than three hours a day and thus, cannot expect minimum wages.

As one of the respondent (manufacturer) remarked,

> "Why should we pay them more? They do not work for more than three to four hours a day. Even this is quite a liberal estimation. During the winter months (from November to February) the number of hours of work put in falls down further to one or two hours per day. The smaller days (during winter) and lack of electricity in the villages mean that the work can only be done during the free hours of morning and early afternoon. Most of these workers anyway are lazy, shirking away from work all the time. Plus, if there is a wedding in the village, you can be rest assured that the work shall cease for an entire month. It is we who have to bear their *nakhras* (tantrums) and pay them even when they do not deserve it."

Such claims typically make for an interesting read when studied in the backdrop of what the workers themselves have to say about the number of hours that they roughly put in and the time period taken by them to get the job done. On being questioned on the number of hours spent, the majority of respondents (more than 70%) stated that embroidery work kept them engaged for a minimum of five hours per day (Table 7.4). Furthermore, more than 41 per cent said that they worked for (or in excess of) seven to eight hours a day (Table 7.4).

In any regulated industry these workers would be legally eligible for minimum wages. Close to thirty-one per cent of the

Table 7.4: Average Number of Hours Spent per day in Chikankari

Number of Hours Devoted to Chikankari per day	*Rural Respondents (in %)*	*Urban Respondents (in %)*	*Total (in %)*
1-2	16.67	0	7.31
3-4	27.78	21.74	24.39
5-6	22.22	34.78	29.27
7-8 and above	33.33	43.48	41.63

Source: Field Survey.

workers admitted, however, of working for less than four hours per day, eight per cent out of which stating that they worked for just one or two hours in a day; contradicting, in turn, the allegations of the *mahajans* which held more or less the entire workforce guilty of the same.

As is clear from Table 7.4, rural workers on an average, work for lesser hours than their urban counterparts. The minimum that any urban worker reported (to be working for) was three hours per day whereas more than sixteen per cent of rural chikankars admitted working for only one or two hours a day. Most of these workers are unskilled embroiderers who perform a few basic, running stitches to earn some extra 'side money'. They treat chikankari only as one of the many different sources of income, along with farming, animal husbandry, grazing etc. Thus, they are 'part-time' workers in the truest sense. Admittedly, the number of hours put in by these rural embroiderers is significantly low, but that can in no way justify the level of wages compensated to the rest of the workers, the majority of whom still work for more than five to six hours per day. The latter are the proper chikankars who work for longer hours as the nature of design they embroider is relatively more complex and time consuming. Yet even they have to often remain content with the low wages dished out by the *mahajans*.

The fact, however, remains that Chikankari being a household industry makes it virtually impossible for anyone to keep a precise count on the number of hours that a chikankar has worked for. The Mahajan's word is as true (or false) as the worker's. Even if we assume that the latter's replies are far more genuine, the very fact that those hours are spent inside one's domestic space (and not in an organized factory or even a workshop) render any claims for wage hikes worthless in the eyes of the law. No labour laws, it must be noted, are applicable for such household based activities.

Simply put, at the end of the day it is the household based nature of this industry (and the way it is defined and understood in legal terms) that is the primary cause of the worker's hardship—in terms of low remunerations for the work put in. A chikankar might be working for more than ten hours a day, but the *Mahajan* can still pay her in an arbitrary manner since legally the *Mahajan* is not bound to compensate the latter with any standard basic pay. The cultural constructions and stereotypes associated with labour performed by women may account for the *Mahajan's* own subjective opinions as to the amount of wages that ought to be given. Yet any such conclusion arrived at would be pointless, if standard regulations (both in terms of pay and working conditions) are made mandatory throughout the industry. Till that occurs, the artisans of Lucknow continue to lead an existence fraught with uncertainties and risks.

It is interesting to compare the monthly wages of the home based workers with the wages of those few that chose to/were allowed to move outside the house and attend the workshops or "centres" run by some of the prominent NGOs operating in the city. The author collected the data from two main organisations—Self-Employed Women's Association (SEWA), Lucknow and Network of Entrepreneurship and Economic Development (NEED).

We have done a comparative analysis of the average monthly income of the artisans working for an NGO as against the average monthly income of those working for private entrepreneurs or *Mahajans*. It was found that the average monthly income of an artisan working for an NGO was Rs. 1,100 while that of a non-NGO worker was Rs. 594. Thus, if a worker was to shift from working for a private *mahajan* run home based production system to being employed by an NGO, she would experience a wage hike of approximately eighty-five per cent. The difference between their monthly household incomes was also quite significant with the households of NGO members earning close to about Rs. 4,336 per month while the average household income of all non-NGO respondents was found to be Rs. 3,322 with a difference of about 31 per cent between the two.

The wage depressing tendencies of household based production systems are clearly evident. Norms of the *'purdah'* restricting women from working in centrally organised workshops thus, only exacerbates the extent of surplus value siphoned off from the already exploited female workers.

Precarious Existences

The living condition of most of these chikankars, not surprisingly then, was found to be in an extremely sorry state. As has been mentioned above, most chikankari workers hail from socially marginalized groups whose members find job opportunities extremely difficult to come by. Human capital development initiatives (for such groups) have been few and far in between with the consequence that a good number of people belonging to such groups lack even the most basic education and skill sets.

The men in the chikankar's household are mainly engaged in labour intensive occupations such as carpentry, brick-laying, rickshaw pulling etc. A sizeable number are also involved in the traditional *"ari zardozi"* craft work of Awadh. In the rural districts surrounding Lucknow, most men are involved in agricultural activities for a considerable portion of the year. They generally work as tenant farmers or as landless agricultural labourers. During the lean periods (in agriculture), they either migrate to the city of Lucknow in search of unskilled work or take up *"ari zardozi"* work within the village itself. The type of occupations taken up by the members of the chikankar's household, as evident, are generally of a low paying, labour intensive nature.

The fact that most of these chikankars hail from significantly large households accentuates the hardships faced by them. The paltry incomes earned by the embroiderers are mostly spent in providing for a large number of dependants. For many chikankars, the husband's incapability to support the household on his own income, in fact, is the primary reason for their wives to enter the chikan industry in the first place. The large household size of the chikankars can be attributed to a host of social, cultural and economic factors. For one, there is lack of proper information available to the chikankars on maternal and child healthcare practices. Extension services provided by the Government (in this regard) can be best summed up as being extremely disappointing and ineffective. Furthermore, the practice of living in large, joint families is quite prominent, not only in the villages but in the city as well. The latter can be attributed to the fact that low income earning opportunities and high accommodation rents in the city often force extended members of the family to live together since it proves to be a far cheaper option. The practical need of having a greater number of economically productive individuals each sharing the burden of household expenditure also gives rise to larger families.

Quite a number of these households fall under the poverty line; despite the fact that at least one member is engaged in a profession (chikankari) for most part of the year. This is not in any way a flattering statement for a profession celebrated as being part of the very heritage and culture of the region.

Most of the Chikankars that the author interviewed lived in dilapidated, overcrowded houses with no proper water or electricity connections. Very few had a functional toilet in their homes. These houses were, moreover, located in the most polluted, unhygienic pockets of the city which have not done their health a great deal of good. Nor has the fact that they have to sit for long hours in the same posture while embroidering the cloth. In fact, more than half of the respondents interviewed complained of severe backache and pains in their joints while most of the elderly Chikankars reported reduced visibility, infection and other eye related problems. Water infection and other gastronomical diseases were also quite common amongst these workers possibly due to the lack of potable water supply and nutritive food.

In a large scale survey conducted by the Giri Institute of Development Studies, Lucknow, on the housing status of chikankari workers (see Table 7.5) it was found that two-thirds of the households surveyed did not have a bathroom/toilet and the families would either use public latrines or defecate on the roadside. Around one-sixth of the houses had the provision of dry latrines which are legally banned in the country. Piped water supply was available in only sixty per cent of the houses. Many did not even have an electricity connection.

Educational Status

The *a priori* market segmentation, to a large extent determines the range and nature of jobs available to any willing worker in the labour market. In *a priori* segmentation, the labour market is split

Table 7.5: Houses having Toilet, Water Connection and Electricity (%)

Category of Workers	*Toilet/ Bathroom*	*Sources of Water*			*Electricity*
		Tap	*Hand Pump*	*No Water Supply at Premises*	
Chikankar	33.33	60.32	3.17	34.92	63.49
Cutting Master	33.33	33.33	50.00	16.67	66.67
Printer	42.86	71.43	7.14	21.43	85.71
Tailor	16.67	16.67	33.33	50.00	50.00
Washerman	33.33	50.00	–	50.00	100.00
Agent	33.33	88.88	11.11	–	66.67

Source: Singh and Jafri (2008).

according to pre-existing demographic criteria such as age, sex or social economic status. Educational status of the worker also acts as an extremely important criterion in this regard.

A higher educational profile normally assures high wage paying employment opportunities to the worker, thus raising the standard of living of the household. The level of education that a chikankar possesses thus, has far-reaching consequences both for her family as well as the kind of work that she does. For one, educated chikankars are more likely to be aware about market conditions, existing wage rates and the manner in which negotiations are to be carried out with the intermediaries without being severely exploited by the same. Higher educational qualification, in turn, opens up far greater employment opportunities both within and outside the industry.

Education, moreover, has a prominent role to play in changing outdated yet dominant social practices that have proved detrimental to the overall welfare of the workers. The *'Purdah'* system, a custom steeped in notions of gender inequality, for one, can only be challenged *via* increased social awareness and rational thought. Limiting the mobility of women outside the home and confining them to the domestic sphere severely affects the opportunity sets that these women ought to be entitled to. What is even more disturbing is the fact that the 'politics of culture' operate in a manner that ensures such ideas are internally imbibed, practiced and upheld by the women themselves. Jeffrey (1979) and Abu-Lughod (1986) have written detailed accounts of how women reproduce the conditions of *purdah*, even in the midst of conflict and debate about the legitimacy of its demands. To break free of such ideological traps, the proper education of the women in the household is imperative.

Education not only enlarges the opportunity set available to the chikankars, it ensures that the standard of living in their household improves as a consequence. Better health care practices, for one, are observed by them. There has been, in fact, a significant positive correlation observed between the levels of education that a mother possesses and the health status of her entire family especially that of her children (Sarna and Shukla, 1994).

Most of the workers interviewed were at best semi-illiterate. With few lucrative jobs that would hire such unskilled workers, they mostly have to fall back on the Chikankari industry to provide them with much needed employment. Out of the total 50 workers interviewed, 42 per cent were illiterate, 14 per cent literate without formal education,19 per cent dropped out after primary school, 14 per cent dropped out after middle school, 9 per cent has secondary level of education and only 2 per cent had gone to university or colleges. Thus, little or no education (for a majority) leaves them unequipped in negotiating with the agents. They are more often than not denied access to even the most basic information related to the market.

The Politics of the *'Purdah'*

It would, however, be an erroneous reading on our part if we were to place the blame entirely upon orthodox cultural maxims without investigating the position of the entrepreneurs and their own interests in seeing to it that these norms (instruments of accumulation) are maintained. The readily available presence of a large number of semi-skilled women workers willing to work only from their homes is not just (as many would like to believe) a supply side consequence dictated by social/cultural mores but is also closely in tune with the demand side schema of things as well. Home based work, in short, works in favour of the employer in more ways than one. For one, the work is not recognized by any labour law; implying that the manufacturers are at complete liberty to decide the terms and conditions of work. They do not have to adhere to minimum wages nor do they have to ensure that certain basic standards regarding the work environment are met. They sub-contract the work *via* intermediaries and thus, are absolved of all responsibilities towards the artisans. As explained above, the physical segregation guarantees that there is no threat of labour union retaliation or any similar discontentment arising from the ranks of the workers. Contracts being oral in nature imply that the manufacturers/agents can hire and fire at will. With such a large quantity of labour supply at their disposal, they can virtually set the wage rates at will with little or no opposition coming their way. As Swallow (1982) points out, in her study of the Indian garment industry, informal "putting out" arrangements are not necessarily irrational and backward barriers to capital so much as the means by which accumulation can take place. Through the agent, the *mahajan* is able to tap labour resources widely distributed in space. Mies discussing the lace workers of Narsapur, remarks:

> "It was obvious that the extreme exploitation of these women about which they all complained was made possible only by the specific organisation of the production process, namely the individualised form of the household industry and, what is more important, the maintenance and propagation of the housewife-ideology, the woman sitting in the house" (1981:491).

The 'politics of *purdah*' deems it necessary for home-based labour ('idle woman sitting in the house' myth) to be maintained so as to appropriate high rates of profit. A standardized, regulated labour market would increase the cost of production by a substantial amount (by way of relatively higher wages). Moreover, they (the entrepreneurs) would then be forced to take cognizance of their employees' basic rights which would involve further costs. A larger number of workers officially employed by a manufacturer shall, in turn re-define the latter's enterprise size, forcing him/her to pay higher taxes to the government.

The politics of the '*Purdah*', operate at multiple levels. The family serves as the primary site of exploitation for most of these woman workers. Unequal gender relations, first and foremost are, visible in the very social practice of physical segregation. The men in the household are the ones who deny women the permission to move outside their homes in search of better employment opportunities. Undoubtedly, a number of women also imbibe such patriarchal notions and contribute towards their maintenance but that has more to do with the inherent hierarchy in power structures (within the family) and as a consequence, in knowledge production and control (within the household) rather than any firm ideological standpoint emanating from their own side. The type and nature of profession that the women of the household are to engage in are decided by the patriarchs. Moreover, the decision as to how the income is to be spent more often than not also lies in the hands of the men.

Not just the family but the community also acts as sites of power with a number of contestations taking place between different groups, parties and individuals; each trying to assert supremacy over the other. These struggles however, need not take place in direct, explicit ways but can also occur in subtle, implicit fashions. For example, the cultural taboos against a Muslim woman directly participating in the labour market are social constructions reinforced in general by the 'community' to which she belongs to. Control over a person's freedom of mobility is not exercised *via* written laws or rules. Rather, cultural associations are implicitly utilized as tools of power. Moreover, under the garb of a larger collective called the community, powerful groups or classes (from within and outside the community) with their own interests at stake raise obstacles in the path of other groups whom they see as competitor. For example, the *mahajans* and agents see it in their own interest if there is a large supply of household bound female labour willing to work for them. However, they would not openly demand for it but use their relative influence in the community to propagate a sense of general maxim that does the work for them. One way of carrying this out is by consciously employing only those women who work from home. Not only does this create a group-occupation nexus within the industry, it acknowledges and contributes (knowingly) towards the upkeep of the laws of segregation and domestication. The upholders of female 'honour' and 'dignity', the men in the community are given an opportunity to restrict female workforce from participating in the labour force (and giving them competition) by stating that "when work is coming home to the Chikankars (courtesy the *Mahajans*) why do they need to go outside?"

Some of the manufacturers/*mahajans* interviewed even managed to sound magnanimous by claiming that it was through their 'acts of goodwill' that poor artisans were able to make at least some money despite staying at home. Just how benevolent their motives really are, is clearly brought out in the data (Table 7.6) given below, which provides us some idea of the amount of surplus siphoned off the artisans. Being a labour intensive process, we observe that the paid wages though very low, still constitute a significant share of the total cost of production especially for low quality products. Share of wages in the final selling price, however, is quite small from about one-third for low quality products to an abysmal one-sixth for chikan products of a higher quality.

Considering that no intensive physical capital (in the form of machines, plants etc.) is required at any stage of the production process, the figures below are indicative of the manner in which the entrepreneurs by virtue of their control over the chain, accumulate value at the expense of the workers involved. It is to be noted that the Table 7.6 presents a cumulative picture of the wages of all the workers involved in the process. The share of wages accruing to Chikankars alone is in fact much lower. When such figures are put in front of the manufacturers, their immediate response is to blame the middleman or *bichoulia* for the pathetic condition of the workers. Their argument is that the intermediaries siphon off most of the income accruing towards the workers.

In fact, it is not only the entrepreneurs even prominent labour economists, social scientists not to mention Government officials and NGO spokespersons, all seem to blame the agents for the backwardness of both the workers and the industry. Most of the

Table 7.6: Share of Wages in Manufacturing Costs and Selling Price of Chikan Products

Item	*Share of Wages in Manufacturing Cost (%)*		*Share of Wages in Selling Price (%)*	
	Lcw Quality	*High Quality*	*Low Quality*	*High Quality*
Gents Kurta	41.43	17.00	34.81	14.18
Saree	49.38	22.47	37.02	18.86
Salwar Kameez	46.50	19.79	39.90	16.85
Sherwani Kurta	44.85	40.17	39.61	30.97
Ladies Top	49.36	37.38	40.45	27.73
Bed Sheets	51.33	35.07	41.30	28.20

Source: Singh and Jafri (2008).

reports published on the subject have recommended that the middlemen need to be taken out of the picture if the twin objectives of growth in the industry and betterment of the workers is to be achieved.

The *Bichoulia*

Not contesting the fact that the middlemen are often guilty of exploiting the workers for their own advantage, any contention that they must at once be removed from the production chain if progress is desired, is myopic in one's understanding of how the industry is structured and how crucial the *bichoulias* are for the industry's stability and survival. Two assumptions are implicit in these recommendations. Firstly, that the *bichoulias* are a homogeneous group with no differentiation amongst them on socio-economic grounds. Secondly, that the *bichoulias* themselves are in some way responsible for the putting out system, i.e. the modern relations of production are a given if only these pre-capitalist 'actors' are removed from the production chain.

What the experts and the researchers do not realize is the fact that the *bichoulias* too are a product of a larger system that thrives on surplus value generated *via* the household system of production. What is further amiss in their understanding is a deeper analysis of the *extra economic* factors that have contributed to the continued persistence of the putting out system in the 21st century.

The argument being made over here is that just by removing agents will not remove the structural conditions that account for the women's exploitation in the first place. These conditions are to be located in certain social and religious customs which are, in turn, fully exploited by the Industry's *mahajans* for their own gains. Removing the agent shall only force the workers to leave the Chikankari craft and move on to a similar household based activity (such as '*ari-zardozi*') where an analogous form of exploitation shall ensue. Neither of the twin objectives (growth with improvement in worker's condition) will be achieved. The very fact that Chikan is a handicraft and is thus not reliant upon intensive physical capital investment can be seen as another major reason for the continued existence of the putting out system within the Industry. Since most of the work is to be performed by hand, it does not make much difference (in terms of quality and time taken) whether the work is being embroidered at the artisan's home or at the manufacturer's workshop. Though there is greater supervision at the workshop, the excessive overhead costs of maintaining a centre, often direct the *mahajan* to put out their work to home-based artisans.

Calls for removing the middlemen, often tend to perceive the *bichoulias* as a homogeneous group of agents amassing wealth at the expense of poor artisans. What is ignored in the process is the fact that there exist differentiations and hierarchies even amongst the *bichoulias* as well. Agents belong to a wide spectrum of socio-economic backgrounds. It must be furthermore remembered that a host of these agents are former female workers who have defied the stringent social norms and have dared to enter the field on their own terms. Forcibly removing the middlemen, would see these *bichoulias* relegated back to the life from which they have tried so hard to escape.

Conclusion

The important thing to be noted is that instead of putting all the blame on the middlemen, it is far more important to realize that the *modus operandi* of a sub-contracting system is such that the individuals located at the lowest end of the organisational pyramidal are inevitably left vulnerable (in terms of wages and security) to those above them. The putting out system as it exists today propagates inequalities and thus, needs to be eventually replaced by a system of production that allows for standardized wages and proper working regulations.

In order to do so, it is essential to recognize the processes of accumulation of social institutions and the manner in which the sexual division of labour and *'purdah'* suspend the artisans in a constant state of penury.

The precarious living conditions in which the artisans of the chikankari industry find themselves in, necessitates that both state and non-state-actors intervene constantly within the existing mode of production and clamour for changes in order to secure steady, sustainable livelihoods for the artisans. A holistic solution is needed that would eventually assist the overall growth of the industry. However, for this to occur, there needs to be a far more intensive State intervention aimed at, (a) combating the very socio-cultural factors that have acted as an impediment to women's mobility, i.e.

the patriarchal structure still existing within society, and (b) regulating the activities of those controlling the production chain and preventing them from siphoning off the maximum surplus value from the system. The fulfilment of the first objective would, in turn, require strategies initiated by the State that galvanise transformative political action amongst the oppressed artisans and which seek to empower them, both within the household and in society in general. Along with other initiatives, greater levels of education, information dissemination and far more stringent protection of rights need to be prioritised. There needs to be moreover, a thorough re-definition of the epistemological approach adopted while classifying labour performed by females.

The State's current position on the above is best described by an official of the Uttar Pradesh Export Corporation Limited (UPECL) who stated:

> "...the industry works according to a system that is set in its place; if the Government tries to tamper with it, the whole industry would collapse. The *bichoulia* has been at the centre of this system. The maximum the Government can do is to open up centres in the vicinity of the Chikankar's house but that too, would be pointless. They just do not want to work away from home."

The above statement demonstrates quite clearly that the State's attitude towards the industry is in general, *status-quoist* by nature. The Government's own interventionist approach takes for granted the importance of the putting out system for the industry's permanence and growth. It is from the very start presumed by State department officials that the industry cannot survive without the home based system of production and thus there is no need to contest/challenge it. Moreover, it is more often than not uncritically assumed that chikankars would only want to/be allowed to work in the present system of home based production and nothing the Government does can change this. Such an approach totally undermines the capacity of the State to bring about social change and ignores the agency of the individuals to partake in transformative political action. It is agreed that the socio-cultural constructions that bind the women artisans to their households are quite deep-rooted and often supported and perpetuated by the women themselves but they are nevertheless constructions and thus contestable. To assume that the situation is bound to remain as it is discounts the aspect of social change altogether. Moreover, what remains ignored in this entire debate is the interest that the *Mahajans* themselves have in perpetuating the putting out system and maintaining the socio-cultural norms that facilitate accumulation. Incidentally, the only complain that the *Mahajans* generally levy against the State is that not enough "support" is provided by the Government which could help them in taking this heritage art form to the entire world.

In the midst of this general disarray, no one even bothers to take notice of the Chikankar as she silently tries to stitch together not only the chikan cloth but also the shredded fragments of her own life and the lives of those around her.

NOTES

1. The '*purdah*' as a dress code has, in fact, been perceived by many feminist scholars as an empowering tool, which offers increased mobility and freedom and at the same time, protection to women living in modern 'disaggregated' societies (Hosseini, 1996; Afary 1998).
2. The *Purdah* is not a social practice exclusive to Muslim women as Mies (1981) and Amin (1997) rightly point out. However, with reference to the Chikankars of Lucknow, it was observed that rules of domestication and segregation were predominantly observed only by the Muslim Chikankars. The non-Muslim artisans were generally allowed greater physical mobility which to an extent explained their lower incidence as well as higher wage rates.
3. The data was collected by the Author during his field visits to Lucknow and its adjoining districts in the summer of 2000. (N = 50)

REFERENCES

Abu-Lughod, L. 1986. *Veiled sentiments: honour and poetry in a Bedouin society*. Berkeley: University of California Press.

Afary, Janet. 1998. "The War against Feminism in the Name of the Almighty: Making Sense of Gender and Muslim Fundamentalism". *Women Living under Muslim Laws: Dossier 21 (February 1999)*: 7–31.

Amin, S. 1997. "The Poverty-Purdah Trap in Rural Bangladesh: Implications for Women's Roles in the Family", *Development and Change*, 28(2): 213-233.

Art and Artifacts of Lucknowi Chikankari. 2007. Textile Industry Opportunities, Ministry of Commerce & Tcunctad Project, WTO Framework, Trade And Globalization Arena, Fibre2fashion.July 25. http://tcunctad.in/publicationDetails.aspx?iden=33\ (accessed on 25th February, 2011).

Baishya, P. 1997. "The Putting out System in Ancient India", *Social Scientist*, 25(7/8):51-56.

Birdwood, G.C. 1880. *The Industrial Arts of India*. London: Chapman and Hall Limited.

Breman, J. 1996. *Footloose Labour: Working in India's Internal Economy*. Cambridge: University Press.

Bythell, D. 1978. *The sweated trades: outwork in nineteenth century Britain*. New York: St. Martin's Press.

Cain, M., R. Khanam, and S. Nahar. 1979. "Class, Patriarchy, and Women's Work in Bangladesh", *Population and Development Review*, *5*(3): 405-38.

Catagay, N., and S. Ozler. 1995. "Feminization of the Labour Force: The Effects of Long-term Development and Structural Adjustment". *World Development*, *23*: 1883-1894.

Dale, A., N. Shaheen, E. Fieldhouse, and V. Kalra. 2002. "The Labour Market Prospects for Pakistani and Bangladeshi Women", *Work, Employment and Society*, 16 (1): 5-25.

Das, M.B. 2005. "Muslim Women's Low Labour Force Participation in India", in Z. Hasan and R. Menon (eds.), *In a Minortiy*. New Delhi: Oxford University Press.

Dixit, R.S. and Ahmed, S. 1991. "Lucknow Mahasagar ke Chikan karyakartaon ki stithi/sthan—Paryavaran pratyakshikaran evum prabhab par ek upagam" (in Hindi). Lucknow: IT College.

Ekatra, 1994. *Situational Analysis of Women Workers in the Chikan Industry and Impact of Organised Interventions*. New Delhi: Author.

Gordon, D. M., R. Edwards, and M. Reich. 1982. *Segmented work, divided workers: the historical transformation of labour in the United States*. Cambridge, Cambridgeshire: Cambridge University Press.

Government of India. 2006. "Social, Economic and Educational Status of the Muslim Community of India" (Sachar Committee Report), Prime Minister's High Level Committee, Cabinet Secretariat, New Delhi.

Haque, R. 2010. "Gender and Nexus of Purdah Culture in Public Policy", *South Asian Studies*, *25*(2): 303-310.

Harriss-White, B. 2003. *India working: Essays on society and economy*. Cambridge: Cambridge University Press.

Hasan, Z. 1994. *Forging identities: gender, communities, and the state*. New Delhi: Kali for Women.

Hosseini, Ziba Mir. 1996. "Women and Politics in post-Khomeini Iran: divorce, veiling and emerging feminist voices", in Afshar, Haleh (ed.), *Women and Politics in Third World*, pp. 144-147. London: Routledge.

Jeffery, P. 1979. *Frogs in a well: Indian women in Purdah*. London: Zed Press.

Jena, P.K. 2007. "Orissan Handicrafts in the age of globalization: Challenges and opportunities". *Orissa Review, LXIV*, 12-16. http://www.orissa.gov.in/e-magazine/Orissareview/nov-2007/engpdf/Pages12-16.pdf (accessed on 21st December, 2009).

Jhabvala, R., and S. Sinha. 2002. "Liberalisation and the Woman Worker", *Economic and Political Weekly*, 37(21): 2037-2044.

Kazi, S. 1999. Muslim Women in India. A Minority Rights Group International Report. http://www.indianmuslims.info/reports_about_indian_muslims/muslim_women_in_ india_seema_kazi.html (accessed on 11th March, 2011).

Kotz, D.M., T. McDonough, and M. Reich. 1994. *Social structures of accumulation: the political economy of growth and crisis*. Cambridge: Cambridge University Press.

Landes, D.S. 1969. *The unbound Prometheus: technological change and industrial development in Western Europe from 1750 to the present*. London: Cambridge University Press.

Lazerson, M. 1995. "A New Phoenix?: Modern Putting-Out in the Modena Knitwear Industry", *Administrative Science Quarterly*, *40*(1): 34-59.

Leys, C. 1996. *The rise and fall of development theory*. London: James Currey.

Littlefield, A. 1978. "Exploitation and the Expansion of Capitalism: The Case of the Hammock Industry of Yucatan". *American Ethnologist*, *5*(3): 495-508.

Marglin, S. 1974. "What do bosses do?: The origins and functions of hierarchy in capitalist production." *Review of Radical Political Economics*, 6: 60-112.

Marx, K. 1977. *Capital: A critique of political economy*. London: Lawrence and Wishart.

Mathur, R. S. 1975. *Report on Chikan Handicraft, Lucknow. Technical Report No. 2*. Lucknow: Giri Institute of Development Studies, Lucknow.

Mies, M. 1981. "Dynamics of Sexual Division of Labour and Capital Accumulation", *Economic and Political Weekly*, *16*(10/12): 487-500.

Nash, J. C. 1993. *Crafts in the world market: the impact of global exchange on Middle American artisans*. Albany, N.Y.: State University of New York Press.

Neve, G.D. 2005. *The everyday politics of labour: working lives in India's informal economy*. Delhi: Social Science Press.

Paine, S. 1989. *Chikan embroidery: the floral whitework of India*. Aylesbury, Bucks, UK: Shire Publications.

Portes, A. 1983. "The Informal Sector: Definition, Controversy, and Relation to National Development." *Review (Fernand Braudel Centre)*, 7(1): 151-174.

Raju, S., and D. Bagchi. 1993. *Women and work in South Asia: Regional patterns and perspectives*. London: Routledge.

Sarna, T., and A. Shukla. 1994. "A Study of Physical-Health and Neuroticism among Women Engaged in the Home-Based Production of Chikan Embroidery", *Social Indicators Research*, 32(2): 179-191.

Scrase, T. J. (2003). "Precarious Production: Globalisation and Artisan Labour in the Third World", *Third World Quarterly*, 24(3): 449-461.

Sethuraman, S.V. 1998. *Gender, informality and poverty: a global review : gender bias in female informal employment and incomes in developing countries*. Geneva: World Bank.

Siddique, P.A. 2005. "Panchayati Raj and Women in Kerala: The Case of Muslims", in Z. Hasan and R. Menon (eds.), *In a Minority*. New Delhi: Oxford University Press.

Singh, A.K. and S.S.A. Jafri. 2008. *Diagnostic study of Chikan Handicraft of Lucknow*. Lucknow: Giri Institute of Development Studies, Lucknow.

Singh, V. 2004. *Romancing with Chikankari*. New Delhi: Tushar Publications.

Standing, G. 1999. "Global feminisation through flexible labour: a theme revisited", *World Development*, 27(3): 583-602.

Swallow, D. 1982. "Production and control in the Indian garment export industry", in E. Goody (ed.), *From craft to industry: The ethnography of proto-industrial cloth production*, pp. 133-165. Cambridge: Cambridge University Press.

Weber, Max. 1981. *General Economic History*, trns. by Frank Knight. New Brunswick N.J. Transaction Books.

Weiss, A. M. 1992. *Walls within walls: life histories of working women in the old city of Lahore*. Boulder: Westview Press.

Wilkinson-Weber, C. M. 1997. "Skill, Dependency, and Differentiation: Artisans and Agents in the Lucknow Embroidery Industry", *Ethnology*, 36(1): 49-65.

8

A Peek into a Town called Malegaon

Saqib Khan

In recent years, towns like Malegaon have come to be seen as a hotbed of communalism. The riots in 2001 and serial bombings in 2006 and 2008 have strengthened such discourses. It has been seen that most discussions on towns like Malegaon tend to just focus on the communal/secular issue while there is very less debate on a range of issues that the people of this city have to face for a decent standard of living. This chapter is based on the author's field studies in the town during June-August 2010. Briefly tracing the historical antecedents of the city and the processes that have led to the strengthening of the communal discourse in recent times, it analyses civic amenities, health infrastructure, powerloom industry, and education and employment scenario in the town. Throwing light on politics in Malegaon, the chapter argues that the political forces and processes that have thrived in the town have failed to address the concerns of the people at large and led to the weakening of the secular fabric. These have gradually acquired a shift towards the 'Right'. The chapter concludes with the argument that Malegaon lacks basic amenities and faces a number of fundamental issues: from education to economy. The town needs a progressive political force that has a double task: working towards improving the overall standard of living of the people and keeping communal forces at bay and building a secular atmosphere.

Tryst with History

Malegaon is a town located in north-western Maharashtra on the Mumbai-Agra National Highway and is part of Nashik district. It is about 280 kms from the state capital Mumbai, and lies at the confluence of Mosam and Girna rivers. According to 2001 census, its total population was 4,09,403 (Census 2001, Basic Data Sheet, Nashik District) and covers a surrounding of about 13 square kms.[1] About 75 per cent of the population consists of Muslims. Most of these are Ansari Muslims—descendants of the weavers who settled here from state like Uttar Pradesh (UP), and whose mother-tongue primarily is Urdu. However, there are also local Maharashtrians called *Dakhani* Muslims (from Deccan) and whose mother-tongue primarily is Marathi. The rest 25 per cent of city's population comprise non-Muslims while Scheduled Castes (SCs) and Scheduled Tribes (STs) comprise about 11 and 12 per cent of the town's population, respectively (Census, 2001).

Officially an administrative block, Malegaon is one of the largest towns in north-west Maharashtra and the second largest city of Nashik district after Nashik city. The demand to convert the Malegaon tehsil into a district has been going on for a long time (Faizee, 2008).

From being a small junction called "Maliwadi" (hamlet of gardens) being a source of employment for many. Malegaon has had its tryst with history. After a fort came up in mid-eighteenth century by a local ruler, the area witnessed immigration of Muslim workers and artisans from northern and western India. The city had its share of colonial history as well. The British capture of the Malegaon fort in first quarter of the nineteenth century saw Muslims from Hyderabad region migrating to the region while the revolt of 1857 saw migration of *momins*[2] from north India, particularly Uttar Pradesh. In the twentieth century, political upheavals in Hyderabad during 1940s and 1950s saw a similar pattern of migration to the city.[3] Over the years, Malegaon became a centre of traditional handloom-weaving while the era of power-looms emerged from 1930s onwards. According to Nashik District Gazetteer 1883, the major chunk of the population was involved in weaving Grey Cloth and Cotton Carpets (Malegaon Municipal Corporation, 2010).

The growing textile industry of the town attracted huge population of migrant workers and the town witnessed an increase in population—from 10,624 in 1881 (Nashik District Gazetteer) to 4,09,403 (Census, 2001). In 2001, Government of Maharashtra decided to convert Malegaon Municipal Council into a Corporation.

Communal Riots, Bomb-Blasts and Effects

Malegaon has witnessed riots as back as 1963 (Hamdani, 2001).[4] The riots of October 2001 were the biggest after the riots of 1992-93 (Engineer, 2002). Apart from the large number of deaths (13), these riots had some other peculiarities: they occurred in a Muslim majority town and rapidly spread to a large number of surrounding villages, a phenomenon seen for the first time on such a large scale in Maharashtra.[5]

The town has witnessed a spate of bombings in the last three-four years. On 8th September, 2006, thirty-one people were killed and over 100 injured in two bomb blasts in the town when people had gathered for Friday prayers (*The Hindu*, 2006). Similarly on 29th September, 2008, during *Ramzan* and *Navratrotsav*, a bomb explosion killed three persons and 101 were injured with various degrees of injuries (Jaffrelot, 2010).

The above happenings in Malegaon have had serious fallouts for the town and its people. Over the years, especially post-2001 riots, there has been a continuing shift to "safer" localities by people of both the communities, with non-Muslim families moving from Muslim localities in the heart of the town to Camp and Sangameshwar (localities across Mosam River on the western part of the town and predominantly inhabited by non-Muslims), and Muslims from these localities to the heart of the town. The formation of municipal wards is an ample proof of this "polarization", with 14 non-Muslim wards on the western part of the town across Mosam River comprising predominantly the non-Muslim population and the rest 58 on the eastern half and the heart of the town comprising predominantly the Muslim population. Thus, to a large extent Muslim-non-Muslim population of the town has come to be divided across the Mosam River which has become a "border": 75 per cent of city's population (Muslims) living on the eastern half of the river while the rest 25 per cent of city's population (non-Muslims) living on the western half in Camp and Sangameshwar wards; though 'pockets' of mixed areas remain on both sides and on the surface harmony and day to day interaction exist in the town, especially due to the textile industry.

Thus, the cumulative effect of the above happenings has been that Malegaon has been dubbed as a "sensitive town" by the media as well as intelligence department (Hafeez, 2010). In fact, along with Bhiwandi and Thane, Malegaon has been declared an 'ultra-sensitive zone' by the Maharashtra Government.[6] This discourse has come to predominate any discussion on Malegaon and there is hardly any talk on standard of living and everyday issues that the people of the town have to face in quest for these, namely civic amenities, housing and toilets, health, powerloom industry, education and employment.

Civic Amenities

For a Corporation town like Malegaon, the standard of basic civic amenities—drinking water pipeline, roads, toilets, drainage, etc. for a large section of population is abysmal. The town is marked by either lack of or inadequate civic amenities. Lack of *pucca* roads, toilets, drainage and cleaning, drinking water pipeline, etc. are issues which the people of the town have to face everyday. Take for example, Ward Nos. 47 and 64. Localities like Haroon Ansari Nagar, Ahmedpura and Datar Nagar (Ward No. 47) and Jamhoor Nagar (Ward No. 64) are marked by absence of even *kutcha* roads, drainage, drinking water pipeline and toilets.

However, it is also true that as far as civic amenities are concerned there exist differences between the "Muslim-concentrated" areas of the town and the "non-Muslim concentrated" ones (Sangameshwar and Camp), with non-Muslim areas having relatively better roads, drainage facility, public toilets, etc. But even within the non-Muslim areas—Sangameshwar and Camp—it can be seen that there are localities that are worse off than the rest of the areas. And there is no surprise that these are the Dalit and tribal localities. For example, Jooni Bhilati and Bhilati, localities inhabited by Bhils, in Ward No. 1 and Ward No. 2 respectively or Panchsheel Nagar, a Dalit locality, in Ward No. 2 are marked by absence of fundamental civic amenities like proper roads, drainage, drinking water pipeline. In fact, in Ward No. 2, roads have come up bypassing localities like Panchsheel Nagar and Bhilati. Despite these differences within the town and between communities, overall the lack of proper civic amenities: roads, drinking water pipeline, drainage, etc. plague almost all the wards of the town.

Housing and Sanitation

About fifty-five per cent of Malegaon's population lives in slums in abysmal conditions.[7] The town is typically characterised by *patra* houses: made of tin sheets and blocks of wood. Rows of such houses just next to each other in narrow lanes with no gap can be found across the city. The dimensions of such houses of Malegaon usually range from 20 × 15 feet or 15 × 12 feet—generally home to two-three families.

In fact, the critical nature of the town's housing question can be gauged from the fact that in some localities people are currently living on old cemeteries. Localities like *Yasin Miyan ka Takiya* in Ward No. 45 (New Mangalwar) and *Bhau Miyan ka Takiya* in Ward No. 52 (Nayapura) have come up on old cemeteries and till date the cemeteries can be seen inside a number of houses in these localities.

Sanitation is another major concern for the people of the town, especially for women. About 60 per cent of households in the city do not have toilets.[8] The public toilets that have been constructed in some of the wards are limited in numbers and are abysmal. Discussions with groups of women in the town revealed that toilets are of the biggest concerns for them as often they are forced to wait for nightfall so that they can use open grounds as lavatories.

Though most of the wards of the town have drainage, these are open and clogged and demand regular cleaning. The localities that lack drainage are in a mess—rainy season or no rainy season. Some of these localities, for example, Salamatabad in Ward No. 63, manage with makeshift 'pits' in front of each house where effluent collects and a family member (usually a female) throws it out when the pit gets filled. The lack of *pucca* roads in these localities adds to the everyday miseries of the people. Thus, inadequate housing, lack of toilets and sanitation, *pucca* roads, proper drainage, etc. are some of the pertinent and everyday concerns for the people of Malegaon.

Health Infrastructure

Given the lack of basic civic amenities and sanitation in Malegaon, diseases like malaria and dengue are a usual menace and it is no surprise to find plethora of private clinics in the town. Here in these clinics, check-up and medication comes as cheap as Rs. 10–Rs. 15, while the hospitals charge around Rs. 100. However, the flourishing of these has to be seen *vis-à-vis* the state of public health infrastructure and facilities in the town. The two government hospitals—Wadia and Ali Akbar—can easily be called a "patient's nightmare" as they are in abysmal condition and lack facilities for X-ray, urine and blood tests, etc. and patients have to rely on private hospitals and doctors.[9] It was only after the 2008 blast that a Civil Hospital, finally, came up in Malegaon. Inaugurated in June 2009 amid much fanfare, the 200-bed Civil Hospital turned out to be a 'hollow gift' for the people of the town. A cursory glance at the three-storied hospital with lush-green lawns from outside does not tell the real story. Though patients are admitted, the hospital has no back-up for electricity, lacks facilities for operations, etc. and patients are either referred to private hospitals/doctors or to other towns like Dhule.[10] Thus, patients have to run from civil and government hospitals to private hospitals/doctors and to other towns like Dhule. Off late, diseases like polio have also surfaced in Malegaon. The two positive cases detected in Maharashtra in 2010 are both from Malegaon.[11]

Powerloom-Led Textile Industry

The powerloom-led textile industry is the mainstay of Malegaon's economy and employment. Traditionally a handloom-weaving centre, the era of powerlooms in the town emerged after mid-1930s and the cloth/textile industry has flourished since then. Malegaon has about one lakh powerlooms installed in about 5000-6000 units—a unit consisting of around 1 to 24 powerlooms. These are mostly located in the residential areas and thus, it is difficult to distinguish between the residential and the industrial areas.

The production consists mainly of cotton grey and man-made fibre grey fabrics (polyester), while a small segment (about 5%) consists of yarn dyed fabrics used for direct clothing purposes—*saree* and *lungi*. On an average the total daily production of the town stands at 1 crore (10 million) metres worth Rs. 12 crore (Faizee, 2008).

A powerloom brings to light the stark realities of life in Malegaon: labourers working for long hours in the deafening sound of looms, no proper ventilation and light, no proper facilities for drinking water and toilets, involvement of women and children in tedious works, load-shedding, etc. Generally, all the workers are casual wage labourers—appointed informally (nothing in writing) by *Seth*[12], working for about 10-12 hours and running about 6-8 looms, no insurance cover, etc. and removal at will by *Seth.*[13] those in cotton-polyester segment are also prone to diseases like asthma and T.B. Wages is on the basis of metre of cloth produced and given weekly (on Friday). The minimum six-hour load-shedding affects the production and wages as well. Wages generally ranges from Rs. 500 to Rs. 700 for one week. Though workers in *saree* segment get wages in the range of Rs. 800 to Rs. 1,000 per week and run generally two looms. They do not form the bulk of the loom-workers as *saree* and *lungi* account for only about 5 per cent of the total production. Thus, the bulk of the loom-workers are those in cotton and polyester segment whose wages range from Rs. 500–Rs. 700 per week. One such loom-worker is Abdul Samad, 62 years of age and a resident of Islampura in Malegaon. He has been a loom-worker for the last 45 years. After working for 12 hours a day and running 8 looms, he gets a weekly wage of Rs. 700.

Though women are involved in various operations inside the powerloom, they also work from their houses and are involved mainly in running the *charkha*[14], filling colour in thread and preparing rolls of the coloured thread. The prepared rolls are then sold locally and these women after labouring whole day generally earn between Rs. 200-500 every month for this. Though the city is replete with such working women, few examples here would suffice. The case of Zulekha Ansari highlights the condition of these women. Sixty-five years of age and a resident of Nayapura in Ward No. 45, Zulekha works on *charkha*: spinning and putting colour into threads for *sarees*. Labouring for a day she manages to earn just Rs. 6 and monthly about Rs. 150. Widows like Shakeela, aged 39 years and a resident of Tehzeeb Nagar in Ward No. 68, have a hard time bringing up four children (all below 18 years). Working on a *charkha*, she also manages to earn Rs. 200 per month and it's no surprise that the family is dependent on community-help for its survival.

Malegaon's powerloom industry is marked by two important contestations: one which is of 'communal/religious' nature and the other which is more 'class-based'. Firstly, the powerloom industry can be said to be broadly divided along communal lines. The weaving units are controlled by Muslims and both owners of the looms as well as powerloom workers predominantly are Muslims. On the other hand, Hindus have control over yarn manufacturing units and trading of cloth. The loom owners (Muslims) buy yarn from Hindus and the prepared cloth is again sold to local Hindu traders. Over the years, there has been an element of contestation as it has been alleged by weavers/loom-owners that yarn manufacturers take advantage of their control over the market to hike yarn prices and get an unfair share of the profit.[15] Secondly, it is the powerloom worker who ultimately bears most of the brunt: whether be it the above highlighted "loom owner (largely Muslims)—yarn manufacturer/cloth trader (largely Hindus) contestation or any inside the powerloom. Any fluctuation in prices, procurement, etc. is always a bad news for the loom workers who survive on meagre wages and work in tough conditions inside the looms. The loom-workers, especially in cotton/polyester, allege that over the years while wages have not increased substantially, the number of looms and the workers have seen an increase. This is evident from the fact that about ten years ago though a loom-worker used to get Rs. 400 per week, he used to run only three looms everyday; now though he gets Rs. 600-700 per week, but has to run about six to eight looms everyday. Thus, there is an irony of Malegaon's powerloom industry that most often the "loom-owner—cloth manufacturer/trader" contestation is emphasised while the "loom owner *vs.* labour" conflict, which is also quite pertinent, isn't highlighted as such. The lack of other economic opportunities in the town means that loom-workers have to continue in their traditional occupation even though it provides them only with a subsistence level of livelihood. Of late, Malegaon has seen the growth of plastic industry and the town has emerged as a centre for reprocessed plastic pipes and market of plastic semi-finished raw materials.

Education and Employment

The overall educational scenario of Malegaon leaves much to be desired for despite literacy rate of 66 per cent (Census, 2001). One of the critical points of this scenario is the condition of Municipal Corporation's Primary Schools. Most of corporation's 115 primary schools are in a dilapidated condition, mid-day meal has not taken off yet and lacking in basic infrastructure—bench, desk, toilets, drinking water, electricity, etc. In schools like Municipal Corporation Urdu Primary School No. 42, Islamabad, leakage of roof made up of tin sheets during heavy rains means that students have to be huddled in one corner of the classroom. In Urdu Primary School Nos. 13 and 88, it's not uncommon to find classrooms filled with water. Defecation in school campus by local residents, open drains adjoining the classrooms, cattle sheds behind school walls, continuous deafening noise of powerlooms are some of the realities that children of these schools have to face everyday. Schools like Urdu Primary School No. 42 use *agarbatti*[16] to avoid smell of adjoining drain and defecation; also schools like Urdu Primary Schools Nos. 14 and 71 use *agarbatti* to ward off smell of animals tied just behind the school compound wall. While smell is warded off by using *agarbatti*, there is no respite from the noise of powerlooms as these are situated well within residential areas. In fact, it has been held that 33 per cent of primary school students are hearing-impaired because of the deafening din of the power-looms.[17]

Of late, the stand off between the Municipal Corporation and state government regarding the proportion of teachers' salary to be paid by each has taken its toll on the teachers.[18] Though delay in salaries has become a regular feature, in 2010 the delay has been for three months (May, June, and July, 2010).[19] Thus, it is not surprising to find primary school teachers of the city engaged in works like tailoring. The shortage of teachers is another area of concern. Till date, according to the corporation, about 100-120 more teachers are required for the primary schools. The 28 Marathi primary schools that are now left in the town are attended mostly by Dalits and tribals and it's not unusual to find one or two teachers running the whole school. Take the case of Marathi Primary School No. 24 in Himmat Nagar and Marathi Primary School No. 1 in Shivaji Nagar—two of the single-teacher schools. The lone teacher, besides teaching, also has to manage other administrative/clerical work of the school. The lack of teachers means that all the classes (1 to 4) are held simultaneously in one room.

Thus, low wages and the resulting standard of life and lack of economic opportunities in the city as discussed above earlier, abysmal condition of Government schools and the absence of any

Government high school/secondary school/college in Malegaon have a cumulative effect on vast number of children in Malegaon. While girls drop out generally after 8th standard, boys get engaged in a number of economic activities. In fact, an appalling aspect of the city's economy is the involvement of children and it won't be an exaggeration to say that there is hardly any work where children cannot be found in the city—powerlooms, hotels, road-side eateries, tea-shops, garage, hawking, selling fruits and vegetables on handcart, etc. In fact Malegaon is one of those places where the boundary between 'education' and 'work' isn't water-tight. This is evident from 'Night Schools' of Malegaon where children who work during the day and attend classes in the night. One of such schools is the Delight Night High School and Junior College in Rehmatabad. Here children who are engaged in various economic activities during the day attend classes after six in the evening.

As for formal employment, teaching is the only avenue for educated youth of Malegaon. The fact that the town does not have any government high and higher secondary school, college or higher educational institute (medical or engineering) worth mentioning, making the matter worse. Thus, dropping out of the school after middle level and joining the informal sector seems to be the way for a large share of Malegaon's population. Besides powerloom that is the backbone of Malegaon's economy, other main economic activities include that of scrap, hotels and eateries, and petty business. And working as a casual wage labourer seems to be the only "destiny" for the people of the town. This is evident from the fact that an adult working in a powerloom earns between Rs. 500-700 every week while a child who in a hotel or road-side eatery gets usually Rs. 50-60 per day after labouring for almost whole day from dawn to dusk. Malegaon, thus, in the words of the Dutch Labour historian Hubert van Wersch who wrote about Bhiwandi, has become a "pool of cheap labour- largely unprotected by laws and largely un-serviced by civic facilities."[20]

Politics

The range of issues plaguing the life of people of Malegaon is explicably linked with the kind of politics the town has witnessed till now and to some extent which has been 'individual/candidate-centric' over the years. Post-independence, the town had its brush with 'socialists' under Haroon Ansari, who won assembly election from Praja Socialist Party (PSP)[21] in fifties (but later switched to Indian National Congress). In 1960s, Nihal Ahmed also won from this party (he later switched to Janata Party).

The town became a bastion of Janata Party[22] and Janata Dal[23] 1977 onwards and any talk of politics in Malegaon starts with one man: Nihal Ahmad. Nihal Ahmad, currently leader of Janata Dal (Secular), started his political career from PSP, but later joined Janata Party and then Janata Dal.[24] He was Janata Party's and later Janata Dal's MLA (Member of Legislative Assembly) in Malegaon from 1977 till 1999 and is widely regarded as having played a key role in keeping Malegaon divided and backward.[25] Regularisation of slums and issuing of ration cards are the only things he is credited with in his long innings of 22 years at the helm of affairs in Malegaon. Raising the issue of wages of powerloom workers but asking the workers to settle down for a meagre increase was a usual antic employed by him. Thus, it's no surprise when loom-workers allege that though Nihal Ahmad always announced that he was with the workers and even mobilized them, he sided with the loom owners every time as he used to work out a 'compromise' with loom-owners that was always inimical to the interests of the workers. With this hindsight it can also be understood that why any attempts to establish unions by loom-workers were always thwarted by political parties in Malegaon.

Nihal Ahmad's antics, however, gradually acquired a communal colour as he indulged in rabble-rousing and fought elections on Hindu-Muslim plank.[26] By mid-1990s, the self-confessed 'socialist' had turned from a man who used to readily join in the *Ganpati aarti* at Malegaon's various *mandals* into a man who kicked at a *Ganpati pandal* because it blocked the way of a *Moharram* procession. Nihal's JD (S) was finally doomed when Sheikh Rashid of Indian National Congress (INC) defeated him in the Assembly Elections of 1999. This, however, didn't deter Nihal much as he took out a 'pro-Laden', an all-Muslim rally that was found to further vitiate the communally charged atmosphere. As if the communal politics was not enough, the winning of a *Dakhni* Muslim candidate (Sheikh Rashid) in 1999 Assembly Election also opened the flood-gates for a *"Dakhni-Momin"* divide in the politics of Malegaon.[27]

The rabble-rousing of politicians in Malegaon meant an easy fodder for parties like Shiv Sena and Bharatiya Janta Party (BJP) who from the mid-1980s played their part in strengthening the communal politics. Ahmed's cynical politics was matched by the Sangh Parivar's equally cynical use of religion in the Hindu-dominated villages adjoining Malegaon. The Vishwa Hindu Parishad's *Dharmayudh* (religious war) against "Islamic and Christian terrorism", and the Sena and Jaanta Raja[28] built-up an atmosphere which resulted in attacks, for the first time in 2001, on the few Muslims living in these villages for generations (Punwani, 2006). The riots of 2001 saw riots in the town and adjoining villages of Malegaon and 13 people lost their lives in mob violence and police firing[29].

One of the important outcomes of 2001 riots was the alienation of Hindus, and to an extent, of the Muslims from the Congress. There has been growing support to the Hindutva ideology leading to political and social polarisation of the communities. In 2001 itself Malegaon was declared a corporation town. The division of 72 municipal wards is an ample proof of the polarisation—58 wards on the eastern half of Mosam River fall in the heart of the city and predominantly Muslim areas, while the rest 14 wards lie west of Mosam River and predominantly Hindu populated areas, mainly Camp and Sangameshwar. The civic election results of 2007 were an indication of people's disenchantment with parties like JD (S) and Congress as newly formed Indian Muslim Congress Party (IMCP) or the *Teesra Mahaz* (Third Front) pulled off a surprise by bagging 27 seats pushing Congress to the second position with 15 seats. IMCP was formed by Mufti Mohammed Ismail—an influential cleric from the town and the party fought elections on the plank of "development".[30] It must also be noted here that out of the 14 wards in predominantly Hindu areas lying east of Mosam River, Shiv Sena won from seven. Also, relatively new : 'right-wing' parties like Maharashtra Navnirman Sena (MNS) also opened its account and won from two wards.[31]

The results of the General Election and State Assembly both held in 2009 were somewhat a more clear indication of the path politics in Malegaon has begun to tread. In the general election held in May, Pratap Sonwane of BJP won the Dhule Constituency that includes Malegaon. Later in the Assembly Election, while Mufti Mohammad Ismail won from the constituency of Malegaon central on the ticket of Jan Swaraj Party (JSP)[32], Malegaon Outer constituency was won by Dada Bhuse[33] of Shiv Sena.

Thus, politics in Malegaon has had a roller-coaster ride from the 'socialist' days of sixties and seventies to heyday of communal and politics of polarization from the eighties and nineties onwards. Two important points must be noted in this regard. Firstly, the discussion in initial part of this chapter on some aspects of the lives of people of the city suggests that over the years local politics has belied any talk about improving the standard of life of the town's population. The lack of basic civic amenities, health facilities and schooling, no avenues for employment, the condition of power-loom industry, etc. are an ample proof that politics in Malegaon has always hinged on factors other than these. Secondly, it has had important repercussion for communal/secular discourse. The communal overtones of politics in Malegaon in eighties and nineties meant that any talk of a secular discourse in times to come was always going to be difficult. This is evident from the broad polarization of both the communities across Mosam river over the years and growth and consolidation of 'right-wing' parties in the town. The history of politics in Malegaon for the last two decades or so, thus, shows that none of the forces have either focused on issues of the people at large or have made attempts towards building of a secular atmosphere in the town. There is also a big question mark that how well the everyday issues of Malegaon's population and secular discourse will be taken up now with the growth of 'right-wing parties', especially Shiv Sena, on the one side, of the town and the coming up of clerics, etc. in politics on other side of the town. It is in this very sense that Malegaon comes up as a somewhat unique case where on one hand the development needs of the town have to be emphasised and on the other, the secular discourse also needs to be built-up.

Conclusion

It can be seen that towns like Malegaon because of history of communal riots and bombings in the recent past have come to be regarded as a "sensitive city" and it is the communal/secular discourse that has come to be associated with such cities. While there is discussion on the above discourse, there is hardly any talk on standard of life of the people and plethora of issues they have to face on a daily basis in the town. The discussion in this chapter shows that Malegaon is a town that lacks basic amenities like housing, toilets, *pucca* roads, drainage, etc. The lack of adequate government health infrastructure adds to the woes of the people of the town. Though powerloom industry is the mainstay of the town's economy, the loom workers are plagued by low wages, abysmal working conditions and are at the receiving end of the both the conflicts—be it with powerloom owners or that between loom-owners and yarn manufacturers/cloth traders. The educational and employment scenario is not without serious problems either. The condition of Government primary schools (both Urdu and Marathi), drop-outs at middle level, employment of children in a number of economic activities, etc. are some of the serious concerns of education in the town. And the lack of any avenues for employment means that powerloom industry is the only future for the youth. The kind of politics and processes witnessed in the town over the years has not helped the cause of the people and has seen a shift towards the 'Right' in recent times. The case of Malegaon indeed poses a challenge and what the town needs is a progressive political force that takes up the everyday issues aimed at improving the overall standard of life of the people of the town as well as keeping communal forces at bay and building a secular atmosphere. Only such an effort can ensure that Malegaon is remembered for things other than riots and bombings.

Aknowledgement: This chapter is based on the author's field-studies in Malegaon during June-August 2010. The author wishes to thank to Abdul Shaban, Noor Alam Azad, Akbar Ali, Rahul Pathak, Awanish Kumar, Aditi Dixit, Inshah Malik, Rahul Sapkal, Abdul Haleem Kidwai and Taha Rauf for their valuable inputs and critical comments.

NOTES

1. The current population of the town is believed to be more than seven lakhs.
2. Weavers.
3. See "A Town called Malegaon", *Indian Express,* 13th September, 2006.
4. The riots of 1963 started off near Jama Masjid of the city over passing of Ganpati procession (Hamdani, 2001).
5. Concerned Citizen's Inquiry Report into Malegaon Riots, 2001.
6. See "Malegaon: the road to perdition", *The Hindu,* 6th September, 2010.

7. See "Malegaon Turns a New Leaf", *Radiance Views Weekly*, 2007.
8. *Ibid*.
9. In fact during the 2001 riots, one of the victims of police firing—Bilkis Bano—was first taken to the government-run Wadia Hospital, but the hospital didn't have any instruments to detect the bullet. So, she was taken to Farhan Hospital (private), but was pronounced dead on arrival. See "Concerned Citizen's Inquiry Report into Malegaon Riots", November 2001.
10. Dhule is a town located about 330 kms north-east of Mumbai and is 50 kms from Malegaon.
11. See "2 in 7 months, polio returns to Malegaon", *Indian Express*, 7th August, 2010.
12. Owner of the powerloom.
13. Owner.
14. Spinning wheel.
15. Concerned Citizen's Inquiry Report into Malegaon Riots, 2001.
16. Incense-sticks.
17. See "Trapped in a blind alley", *Indian Express*, 23rd October, 2006.
18. Currently the share of state government and corporation in teachers' salary is 50:50. But the corporation wants this share to be 80:20 in its favour.
19. As of August 2010.
20. See "A Town called Malegaon", *Indian Express*, 13th September, 2006. Hubert W.M. van Wersch (1992) is best known for his magnum opus *Bombay Textile Strike: 1982-83* (Bombay: Oxford University Press).
21. It was formed in 1952 after the merger of Kisan Mazdoor Party and the Socialist Party.
22. One of the political parties opposed to the Indian Emergency (1975-1977). It formed the first non-Congress Government in the Centre in 1977.
23. It was formed in 1988 by the merger of the Janata Party with other smaller parties. It came to power in the Centre in 1989 under the leadership of V.P. Singh.
24. It was formed in July 1999 by the split of Janata Dal party and is led by former Prime Minister of India—H.D. Deve Gowda.
25. See Concerned Citizen's Inquiry Report into Malegaon Riots, 2001.
26. Popularly it is referred as "H-M" in Malegaon.
27. Though predominantly inhabited by Ansari Muslims, Malegaon also has local Maharashtrians called Dakhani Muslims (those from Deccan). Generally, there is no inter-marriage between the two.
28. Jaanta Raja is an organization set-up by some members of Shiv Sena inspired by Sena's Anand Dighe. In Malegaon area, Jaanta Raja was set-up by Dada Bhuse—a former civil engineer and

builder and businessman by profession. Its members are said to be engaged in "social service" and settling petty problems of villagers. During 2001 riots, the organisation was involved in attacks on Muslims.

29. Out of 13 dead, one was a Hindu; four Muslims were killed by Hindu mobs and eight died in police firing. See Concerned Citizen's Inquiry Report into Malegaon Riots, 2001.
30. The IMCP is a party of ulemas (religious scholars) and "non-political" community leaders headed by chief Imam of Jama Masjid of the city: Mufti Mohammad Ismail.
31. Founded in March 2006 in Mumbai as a break-away faction of Shiv Sena, MNS is a right-wing regional party based in Maharashtra.
32. Though Mufti Mohammad Ismail had established IMCP before civic elections and his party did reasonably well, he stood from JSP in Assembly Election.
33. He is a former leader of Jaanta Raja. During the 2001 riots, Bhuse, along with Suresh Gawli of Shiv Sena, was arrested for the murder of ex-councillor Khalil Ahmad.

REFERENCES

Engineer, Asghar Ali. 2002. "Malegaon Riots—A New Era of Violence", *PUCL Bulletin*, January, Available at http://www.pucl.org/Topics/Religion-communalism/2002/malegaon.htm Accessed on 31st July, 2010.

Faizee, Aleem. 2008. "Malegaon: Not The One You Know", *Ummid*, September, Available at http://www.ummid.com/malegaon/malegaon_not_the_one_you_know.htm Accessed on 1st October, 2010.

Govt. of India. "Basic Data Sheet, District Nashik, Maharashtra", Census 2001.

Govt. of India. "Population, population in the age group of 0-6 and literates by sex—Cities/Towns 2001", Census 2001, Available at http://web.archive.org/web/20040616075334/http://www.censusindia.net/results/town.php? stad=A&state5=999 Accessed, 20th August, 2010.

Govt. of Maharashtra. 2007. Maharashtra Government Gazette (Extraordinary), 30th May.

Hafeez, Mateen. 2010. "Policing in Malegaon", Blog: *Malegaon—My Home Town*, 23rd August, Available at http://mateen-malegaon.blogspot.com/2010/08/policing-in-malegaon.html Accessed, 1st September, 2010

Hamdani, Shakeel Ahmed. 2001. *Malegaon Kal Aur Aaj: Tehzeeb aur Saqafat ke Aaine Mein* (in Urdu). Malegaon: Hamdani Publication.

Hameed, Sayeda. 2006. "Trapped in a blind alley", *Indian Express*, 24th October, Available at http://www.indianexpress.com/news/trapped-in-a-blind-alley/15174/2 Accessed on 28th September, 2010.

Jaffrelot, Christophe. 2010. "Abhinav Bharat, the Malegaon Blast and Hindu Nationalism: Resisting and Emulating Islamist Terrorism", Chargesheets and Miscellaneous documents regarding Malegaon Case, *Economic and Political Weekly* (*EPW*), Vol. 45, No. 36, pp. 51-58.

Malegaon Municipal Corporation: "A Brief History of the Civic Body", Available at http://www.malegaoncorporation.com/main_pages/history_of_corporation.htm Accessed on 1st August, 2010.

Mascarenhas, Anuradha. 2010. "2 in 7 months, polio returns to Malegaon", *Indian Express*, 7th August, Available at http://www.indianexpress. com/news/2-in-7-months-polio-returns-to-malegaon/657168/0 Accessed on 30th August, 2010.

Philipose, Pamela. 2006. "A Town called Malegaon", *Indian Express*, 13th September, Available at http://www.indianexpress.com/news/a-town-called-malegaon/12558/0 Accessed on 5th August, 2010.

Punwani, Jyoti. 2006. "The man who could have built Malegaon—but did not", Blog: *Malegaon—My Home Town*, Available at http://mateen-malegaon.blogspot.com/2010/06/man-who-could-have-built-malegaonbut.html Accessed on 20th September, 2010.

Radiance Views Weekly. 2007. "Malegaon Turns a New Leaf", Available at http://www.radianceweekly.com/62/574/RAJASTHAN-ON-THE-BOIL-Meenas-Gujjars-left-to-fight-it-out/2007-06-10/Inside-India/Story-Detail/62/560/RAJASTHAN-ON-THE-BOIL-Meenas-Gujjars-left-to-fight-it-out/2007-06-10/Editorial/Story-Detail/malegaon-turns-a-new-leaf.html Accessed on 14th September, 2010.

Swami, Praveen and Anupama Katakam. 2006. "Malegaon: the road to perdition", *The Hindu*, 9th September, Available at http://www.hinduonnet.com/thehindu/2006/09/09/stories/2006090907431100.htm Accessed on 30th September, 2010.

The Hindu. 2006. "31 killed, 100 injured in Malegaon blasts", 9th September, Available at http://www.hindu.com/2006/09/09/stories/2006090916980100.htm Accessed on 31st July, 2010.

Wersch, Hubert W. M. Van. 1992. *Bombay Textile Strike, 1982-83*. Bombay: Oxford University Press.

9

Building Inter-Community Harmony: Interventions and Practices in Mumbai

Ruchira Goswami

History has etched out that the rise of one ruler means the fall of another prior unfit ruler who could not stand the test of 'survival of the fittest'. Soldiers were sacrificed in the name of gaining power in order to reach out to the uncharted territories that the world had to offer. Now with evolution, the modern man prides himself over following the matured path of fighting battles for the coveted position of the ruler—that is, *via* democratic elections. However, does the modern democratic leader still use the common man as pawn to meet its goal? Have such battles or competition degenerated into riots?

Reports have repeatedly indicated that riots emerge in a context of competition which is either over political power and/or livelihood resources. Most often riots are perceived as spontaneous occurrences that arise out of petty quarrels. One can either view such incidents as a last straw that gave way to the passive aggressive relationship that exist between the two communities or such incidents are triggered by the competitive forces. In any case, the former as well as the latter explanations cannot be seen in isolation.

Economics and politics are part and parcel of social life. So does that also mean that instances of riots are inevitable and unavoidable? Further, 'what happens to the social relations in a community that has witnessed the wrath of communal riots?' are questions that many academicians side step in the discourse on communalism.

Shourie (1987) holds that the impact of such frequent riots due to the continued inaction on the part of States results in communities being seized with the fever of revenge and such forms of reciprocal terror is equivalent to 'mutual suicide.' According to Mander (2009), there are some who still argue that the survivors of the bloody partition are yet to experience closure and their families are still in agony and feel the loss of their loved ones and their homeland; and all this covert violence and injustice rages unseen in the hearts and minds of the people. That is why, in order to cease this continuously dangerous deterioration of relations between communities, processes of reconciliation and harmony need to take strong roots.

The word 'harmony' in its archaic form implies a musical note that produces a pleasing effect. Through the ages when this term started being applied to express communal relations, one perhaps yearned for pleasing human interactions in the community. The findings of Siobhán McEvoy's (2000) study indicate that peace is not just seen as mere absence of riots. Now, in a country like India, the assumption that any debate over religious practices, and arrest of communal politicians are likely to become inflammatory is widely prevalent. This excuse is used by the general population as well as the government in order to justify its refusal to arrest a politician delivering hate speeches due to the fear of the arrest leading to a riot. Thus, instances when streets become empty at the time of declaration of Ayodhya issue verdict, point out the distrust, fear and anxiety that still reside amongst people.

Moreover, in the same study, school students equated peace with descriptions of equality, of not being treated as second class citizens. In fact, more than the absence of war, peace was envisioned as practical parity of esteem for all citizens (McEvoy, 2000). Therefore, keeping in mind the findings of Sachar Committee Report and Siobhán McEvoy's study, one can conclude that there is peace only from the dominant group's perspective. The marginalized and stereotyped community of Muslims is yet to achieve peace in spite of being citizens of the same nation.

In 1950, the leaders presented the people of India with a Constitution that saw through the religions and since then, the nation has been endeavouring to achieve secularism. The concept of secularism is essentially stating that religion is one's private affair, whereas a secular way of living is the Stately affair. However, in a country where religious books outsell other books and it is religious pilgrimages that attract congregations in millions, the strength of secularism is constantly under stress (Madan, 1992: 397).

In opposition to the principle of secularism, is the thriving process of 'othering' practiced by an in-group against the members of the out-group in the form of holding stereotypes and biases against the latter. This in turn adds up to create roots of prejudice that makes the situation vulnerable to flame riots. Research in social psychology states that prejudice is highly resistant to direct attempts to changing it. This is of no surprise as theories under the larger 'symptom theory' explain how prejudice helps the individual in solving some basic inner conflicts, tension or anxiety. In addition, with reference to the field of cognitive psychology, human beings are often equated to being cognitive misers (Williams and Giles, 1992: 474). Humans look for short cuts in mental processes, that is, they engage in selective attention and perception. One is more likely to pay attention to those incoming information that match the beliefs that are already held by the individual (Weiner, 2003: 313). Ultimately, it is the information that one mentally attends to is what is likely to determine one's decision-making and the consequential behaviour.

Therefore, by residing in a society where engaging in passive aggressive behaviour against the out-group is rewarded and the human mind has a natural tendency to make use of the fundamental components of prejudice, stereotyping and biases, it makes it even more important to understand the content, methodology as well as the medium and frequency of implementation of interventions to promote communal harmony. This is because at the end one has to accept the target group with all its mental processes that fuel stereotypes.

Method

First and foremost, the concept of "prejudice" that is the key component of building the conception of the "other" and the "out-group" cannot be gauged through objective numerical methods. Keeping such an argument in mind, this study resorted to using the qualitative method of in-depth interviews as a tool for data collection with the intention to further understand the work of harmony building interventions in areas prone to communal tensions in Mumbai. In order to get a holistic understanding of the issue of harmony building interventions in the current social development sector of Mumbai and beyond, the sample size included four interventionists who work on communal harmony building interventions (Cases A, B, C & D), four social workers engaged in Non-Governmental Organizations (NGOs) working in areas of Mumbai having a history of communal violence (Cases I, J, K & L) and lastly four participants who engage with a particular Non-Governmental Organization that works on communal harmony building interventions (Cases E, F, G & H). Due to ethical concerns and request from the interventionists/interviewees, they have been referred with code names rather than actual names.

The interviewees mostly resided or worked in slums of Jogeshwari East, Kherwadi and Behrumpada which have not only witnessed the wrath of 1992-93 riots, but over time have become more pronounced neighbourhoods clearly associated with respect to specific religions. There hasn't been any outbreak of riots in Jogeshwari East, Kherwadi or Behrumpada since 1992-93, but one cannot help but speculate over the quality of relations existing between the religious communities within these localities with a history of communal tension.

Hindu-Muslim Riots and Participants

In order to understand 'peace building', one has to first understand and identify the various factors that ultimately result in making certain people participate in riots. What is also equally and even more important to understand is the processes that can have an influence on the large proportion of the dominant community that gives its silent social sanction to such brutalism in the name of religion.

Identity politics is held responsible by all the respondents in this study as the prime cause of riots. That is, when a certain religious community is mobilized on political lines for a continuous period and this is done by demonizing the 'other' through the use of alienating stereotypes. But what is also worth exploring is how the identity of religion can be easily manipulated to create riots. After each instance of riot it is popularly held that such violence is a result of religion being exploited by violent people. One easily interprets the riots as "misdeeds perpetrated by a gang of criminals...of unscrupulous politicians in league with a handful of fanatical religious groups" (Banerjee, 2002: 1183). But, are there certain elements within the overpowering institution of religion which makes it that much more feasible for being exploited by aggressive political ambitions? Case A says:

> "Religion is associated with faith and faith is a central aspect in the occurrence of riots. By killing the other one thinks or feels that they are defending their faith."

According to Carroll Saussy, religious faith has a connection with self-esteem as after handing over one's problems to God one is determined that they would pull through any problem (Saussy, 1993: 364). In addition, Helen Joseph went on to state that due to the economic struggles faced by a common man in liberalized economy, his disappointments and frustrations translate into feelings of hopelessness and meaninglessness (Joseph, 2000: 665), all of which make the person even more dependent on religion for faith and thus, establish a sense of stability and control over his life just so that his anxieties are in check.

It is generally held that after 1992–93 Mumbai riots, discrimination by majority community against the minority community has escalated, leading to latter's migration to areas of its numerically higher concentration. Despite the sense of safety in numbers, somewhere such a move is definitely not in favour of leading up to communal harmony. Case C brought out an association between ghettoization and the increasing trends in migration from rural to urban areas in search of a better livelihood. The resultant competition over jobs becomes very high and thus, the already impoverished people who are torn away from their roots are likely to develop a sense of uncertainty and insecurity. And soon enough, issues of development such as water and sanitation get communalized through debates such as 'through which area does the facility of water supply pass?'

On the other hand, Case B tried to answer the question of why identity politics breaks out into a riot against a marginalized community with the help of an economic perspective. It is the lower middle class that gets most easily mobilized and this is so because within the current competitive market, they are struggling for jobs or are 'job seekers'. However, those who fall among the power houses of giving jobs, may also have their set of prejudices against the minorities, but, they don't engage in riots because of the fact that they don't have to fight for jobs, rather they are the 'job givers.'

Thus, in all the above reasons what comes out is that a common man is continuously struggling to attain a sense of control over circumstances that determine his life, the root cause of which is 'competition'. Due to this interventionist D is not in agreement with interventions such as organization of cricket matches on festive occasions to promote communal harmony, as he believes that by doing so one is introducing the phenomenon of 'competition.' According to him competition leads to a sense of deprivation and discrimination. This is why he does not use the medium of games to bring about peace, harmony and equality. Instead, he prefers giving all the participants small gifts so that everybody feels that they had participated there. Such belief is consistent with Helen Johnson's views about how the current value system of the Indian society is based on the 'win-lose paradigm' which can also play a role in triggering riots (as cited in Joseph, 2000: 668).

The process of socialization is continuously influencing as well as being influenced by social mediums of education and media. That is why, the discourses being delivered by these mediums become important determinants of one's identity. According to Case C, our education system is creating masses who are not made to engage in reflection, as it is a system that encourages all of us to "learn by heart" (*"hum sabna to karte hai"*). This tendency to be non-reflective becomes breeding grounds for succumbing to rumours and impulsive behaviour.

The government, can also be held responsible in various ways for their inability to deal with riots even though they are in a capacity to do so. Case B brought out how due to the purposive negligence or hesitation on the part of the government riots happen. He went onto say that according to V.N. Rai (a police officer), *"If any riot prolongs beyond 24 hours it means state is complacent."*

Hate speeches, spreading of rumours are all starting points of riots. If the laws framed for these offences are implemented by the State then there will be no riot. This act of arrest would imply that no one has the power to escape from the reach of the law and thus, the power of the dominant community would also be effectively reduced. Case B assertes as well as critiques the State's frequent stand on such an issue by stating,

> "Justification the government gives is if we arrest there will be more riots. That is not an excuse because even if there are more riots the primary duty of the state is to uphold the law."

All the participants also blame the government for its role in causing and promoting communal disharmony and riots. They point out the tendency of the government to reinforce the various stereotypes that are associated with Muslim community through rampant arrest of Muslims due to hasty investigations. Case F says: *"Kahi pe kuchh ho gaya toh – sidha ilzaam Muslims par. Yeh ghalat hai"* (If any terrorist activity happens anywhere—Muslims are directly blamed. This is wrong).

Differences in the Perspective of Participants

What prominently came out is that all the interventionists as well as the participants realize the politics of riots. These politicization takes shapes through the manipulation of religious symbols that enable the politicians to achieve power, profit and publicity. However, the interventionists happened to be more aware of the various means of politicization, whereas the participants had a more intuitive understanding of how politicians use the community.

Since the participants hailed from an area that has seen the wrath of riots, their responses of why riots happen emphasized the nature of the rioters which causes those people to engage in rioting. The responses ranged from the fact that the rioters don't know and understand what they are doing, they are non-reflective about their decisions to the tendency that such people follow just one perspective whilst making a decision as opposed to being open to many more perspectives. The other explanations focused more on the current socio-economic scenario of the nation. For instance, according to them these rioters were jobless and uneducated which is why they succumb to the temptation of money offered to them in return of their participation in riots. Therefore, one can also draw out that being uneducated makes it difficult for one to attain a decent job. And probably it is this 'cut throat' economic competition that not only determines who would be the winner or the looser, but also who would survive, perish or get marginalized.

The Perspectives of Non-Interventionists

In case of the respondents working in the development sector but not specifically in the field of communal harmony, it was found that some of them harped on each of the points brought up by the interventionists—politicization, inactivity of the government, etc. However, what was most pronounced was the discomfort of Cases F and G who believed that there was no point in exploring the space of communal harmony in the welfare sector as the people want to move on and their needs are now different. But all the four non-interventionists in their own way brought out numerous communal processes that are still prevalent in the communities—be it the process of othering or the underlying distrust, hatred, power issues, politicization or even joblessness amongst the youth. The interventionists linked these various factors in order to show how a riot occurs, whereas the non-interventionists were listing the factors down without any direction as to indicate how these factors affect one another for a riot to break out. However, it was very interesting to see how Case E summed up the perspective with which welfare sector must deal with communal harmony. According to her, "even though there is no communal conflict at present *one requires the presence of an omnipresent entity committed to communal harmony*", thus, indicating the underlying insecurity running amongst the Indian population.

Forms of Interventions for Communal Harmony and their Impacts

Over the years, certain organizations or people *per se* have been engaging in activities so as to alter or prevent the course of events that have the strength to lead up to a communal riot. The genesis of the term "intervention" can be traced down to the action word of 'to intervene' that is to act in order to make some modification, addition or subtraction with the sole purpose of improving the situation concerned. However, in order to understand "intervention" as a whole, one cannot just limit oneself to the action of only performing the task *per se* that is likely to lead to a social change. In the case of communal harmony, one needs to understand its interventions in totality. In other words, one needs to see keeping in mind factors such as the intention with which the intervention is being conducted, the content that is being imparted, the mediums that are used to deliver the content, the processes that are further incorporated within these mediums, the kind of target group one is catering to, the way the target group responds to the intervention and the way the interventionist is perceived by the participants. In India, communal harmony is yet to enter the dominant mainstream understanding of social issues. Thus, the understandings and misunderstandings those are associated with an issue that one is trying to intervene in, also influences the intervention's sustainability, degree of reach out and its scope of leading up to the desired social change.

Target Groups

Keeping in mind that 'communal harmony' is not seen as one of the prominent social issues in the development sector, unavailability of sufficient resources is a frequently seen as impediment in the implementation of communal harmony interventions. Thus, the target group that one reaches out to when dealing with such an issue is intentionally narrowed down. Cases A and B's target groups include people from diverse backgrounds. But they particularly target those groups which directly intervene in the society or the youth who are *"most receptive, in spite of being at the forefront of riots."* Other target groups include school children, college students, teachers, social workers, bureaucrats, police training schools, teachers training colleges, law schools, NGOs, and political workers.

Case A tries to prioritize the kind of target groups he is reaching out to, which is why he actively seeks an opportunity to do programmes with social workers, media and political workers. Thus, when it comes to even taking a session there is a strategy in his mind, through which he hopes bringing about change in such minds that would have the maximum possibility of impacting a chain of people. In other words, interventionists are trying to sensitize the large majority of silent spectators who give their consent by not raising their voices against communal forces.

Case A very often engages in writing numerous books of varying degree of complexity, so as to cater to readers ranging from academicians to children. In fact, one of the *non-interventionists* happened to state that a certain book changed her entire perspective towards the occurrence of riots as through that book she was introduced to the whole concept of politicization and production of riots. The extent to which general books as well as school books can have an impact in changing people's perception towards an issue can be seen in *Romila Thapar's* article where she exposes us to the extent of dissent on the part of the political forces against the syllabus of NCERT books that contained information which went against their rhetoric of right wing nationalism and Hinduism. For instance, the right wing Hinduism has problems with facts such as 'Aryans along with venerating cows, had its flesh on ritual occasions or when honouring guests' (Thapar, 2009: 95).

Cases A and B use the medium of workshops, lectures and public meetings in order to prevent the spread of communalism. These sessions are specifically interactional in nature. For instance, Case A starts off the sessions by throwing open questions regarding what type of ideas do people have with reference to any topic related to communalism and in this way, he builds the session based on what people think. Case A believes that all this hatred is a result of the manipulation of history that is primarily done through school books and media. Peter N. Stearns (2011) has explained the crucial role of history in defining and enabling comprehension of the factors from the past that cause social change. Such information in turn has an impact on one's understanding of the societal functioning. This makes each one of us a human being with our own set of historical baggage which also influences the way we perceive one another. Such an understanding or belief translates into Case A's emphasis on tracing down history that exists in the minds of people. And it is here on he introduces new concepts, arguments or information that must have been left out in the discourse they were exposed to. In addition, this two-way form of communication is done so as to maintain the attention of the audience as well as make the maximum possible impact within the limited time. If time permits, he screens a movie first and then follows it up with a discussion. He considers films as powerful medium and whenever it's possible he tries to supplement his interactions around films.

Religious Background of the Interventionists

What is also interesting to explore is that whether the religious background of the interventionists has an impact on the target groups and whether the interventionists are conscious of their own religious background when dealing with a target groups.

Case A somewhere felt that as a Hindu interventionist whenever he spoke to a Muslim audience, he felt that the Muslim audience felt reassured to find a Hindu person talking in favour of minority rights. In addition Case A also believed that the whole process of demonization has an impact on people's 'self-worth' due to which they feel guilty and his clarification of history enables them to understand that their forefathers were not as evil as they are portrayed to be in the Indian communal discourse. Such processes of 'othering' not only makes the community more and more inward looking, but also makes them imbibe the feeling that they are in this position because they deserve to be. Such an observation can be corroborated by *Helen Joseph's* experience of organizing communal harmony programmes with the youth. Her team found that at times the minority group's understanding of themselves was based on the stereotypes and myths, the people of the majority groups spread against them. Hence, working on dispelling myths and stereotypes becomes even more important for the victimized community as well (Joseph, 2000: 670). Thus, one can speculate that in this way it leads to a generation of Muslim Community that is subsumed by learned helplessness and in-action against discrimination—all of which are deterrents in achieving one's due rights.

Another interventionist, Case C hails from the Hindu community, but she has considerable knowledge of both Hinduism as well as Islam. She believes that it is this that probably enables others to see her as neutral. Moreover, she is also seen as a person who neutralizes differences. The youth see her as closer to their age and are thus likely to get more motivated by her. Case C also believes that one of the strengths of her NGO is that there are people from different backgrounds. Incidentally all the women working in her NGO are Hindus and the men are Muslims.

Reaction of the Target Group

At the time of the communal harmony session, different people in the audience get affected by the information being received differently. There are those who are easily convinced, while there are also those who begin to question their beliefs. Some remain indifferent and there also some who become hostile. People who become hostile engage in counter-questioning. In such situations, all the interventionists respond to them with reason, logic and many examples. In fact, all four of them even prefer people being open with their hostility, so that there is a scope for bringing the falsely held beliefs to the fore for further discussion and critical analysis. As Case C says:

> "..till you have the space to disagree, you will know how to deal with conflicts. You will always play it safe and agree to things that you don't believe in. If the place is safe and you dialogue it out, you thrash it out. You will go back with a renewed understanding."

Weekly Group Sessions with the Same Participants

The intervention adopted by the NGO in Jogeshwari East where Case C is working is slightly different from Cases A and B. It is striving to bring about a meaningful change in the society through interventions which enable the youth to think and reflect critically and accept and appreciate diversity.

This NGO works within a community that has been affected by riots. The experience of working with riot-stricken community has led Case C to believe that communalism does not just end with riots; rather it is a beginning of a more complex form of communalism where distrust and fear subsume the community spirit. Case C says:

> "By just to deploy police force or military does not mean dealing with riots. For the time being you may be able to separate people and say don't fight otherwise we will punish you. But, it doesn't end there and we are seeing it day in and day out. Aren't we all living in the shadow of our history?"

Case C does not believe in bringing in issue of communalism until and unless that group has been through a series of sessions which instils trust on one another as a result of which a space is formed wherein they can share openly. Such safety enables them to share their opinions on the issue of communalism openly, without the fear of being judged or without the fear of being politically correct. In this way, the entire group understands the context of that person and then it's a movement from antipathy/sympathy to empathy which is very important. Such a nuanced understanding about how to get suppressed emotions about riot experiences out of the individual comes from her experience of working with individuals who have gone through the process of group sessions as a result of which they were able to share their experiences and feelings about Mumbai riots (1992-93) even after years. What is also important here is that such platforms where self-disclosure is facilitated, enables the inter-religious group interaction to become more personalized so as to ultimately build a sense of commonality that goes beyond the physical appearances, thereby reducing prejudice (Dovidio and Gaertner, 1999: 103).

Case C believes that their form of intervention is a process which is not tangible like distributing notebooks or giving free medicines. What they intend to do through their weekly group sessions is to build communities, competencies and perspective among the youth that is not only well intentioned, but also sustainable.

Another annual event that the NGO, where Case C works, organizes and looks forward to is that of Annual Inter School Sport Meet where children from English, Gujarati, Urdu, Marathi and Hindi medium schools come together. In this way, there is interaction across children coming from different languages, religion, and region. Somewhere it also cuts across class because there are children from both the English and vernacular medium schools. Whilst organizing the events Case C and her co-workers realized that these schools don't communicate with each other and this may also be because they are at times located in different religious ghettoes. As a step closer to their mission for the last two Sports' Meet they have asked the physical education teachers of the respective schools to be part of the planning and organization process. The teachers not only spoke for the first time, but spoke about issues related to inputs and debates. The game of football being a great team game was also incorporated. The teams were mixed even in terms of gender and the students were wearing uniforms of their respective schools. So, when a certain team won, the trophy was held up by students belonging to different schools who were still united together as a team.

Impact of the Harmony Building Interventions

What was most amusing and intriguing about the participants who have been associated with the NGO where Case C is working, is that they all spoke of values that they had imbibed through their association with the NGO, but they could not connect the very existence of those values and characteristics to achieving the larger goal of removing stereotypes and prejudices. The four participants repeatedly stated the importance of reflecting on one's thoughts before taking a decision and attempting to understand the perspectives held by others. These two ideologies and abilities can prevent someone from impulsively saying or doing something that can later impact communal relations in a negative way. The three Muslim participants repeatedly conveyed their gratitude to the NGO for improving their communication skills, boosting their confidence and self-esteem as a result of which they were now able to express and communicate what they felt from within. A recent study by Sherman and Allen (2011) found that people, who feel badly about themselves, are more likely to show bias against people who are different. By degenerating others, one feels better about themselves. Thus, one can connect high esteem with more security and less likelihood of engaging in prejudiced behaviour.

Such confidence and self-esteem translates into making the participants more assertive. In this way, the participants can engage with the governmental authorities with rights based perspective and expect performance on the part of the authorities who are accountable to them. Thus, this basic characteristic of good communication can actually serve the purpose of resisting learned helplessness that often gets imbibed in a community which has been marginalized at the wider forum. This heightened feeling of self-esteem and confidence can also indicate that the participants have a sense of control over their decisions which can enable a person from not getting easily swayed by what others say.

The Contact Hypothesis by Byrnes and Kiger states that conditions such as common meaningful goals, equal status contact and diverse inter-group interactions may result in reducing prejudice (as quoted in Williams and Giles, 1992: 475). Thus, the work of Case C's NGO not only creates such an environment where the above conditions are met, but also instils in their participants the right to be treated equally. This increased confidence in turn also has an impact at the time of inter-religious interactions as findings of studies conducted by Hamilton and Troiler (1986) argue that affective and motivational factors influence information processing by increasing the likelihood of selective attention and heuristic shortcuts. This implies that anxieties at the time of inter-religious interactions leads to further chances of engaging in stereotyping or only picking up information that reinforce the existing stereotypes.

However, the point raised by the one interviewed Hindu participant about being cautious when talking about the issue of 'communalism and communal harmony' in front of the larger public, again makes one ponder over the extent to which social development sector can work on this issue in a society that cringes at the thought of discussing an issue of such importance.

The current affairs of the social development sector are such that the work on the issue of communal harmony becomes even more difficult. Cases A and B were particularly unhappy with the quantity of the work that was being done. Even though, they were going out of their way to reach out to as many people as possible across the nation, they still felt that the numbers were very low compared to the enormity of problem.

Cases C and D spoke about their woes over difficulty in getting funds when working on the issue of communal harmony. Then in such kind of work it is difficult to measure or even document impact, which makes it even harder for them to get minimum funds. Owing to these inherent drawbacks in their choice of issue to intervene in, they draft proposals for funding in such a way that that organization gets funds for some other activities but side by side they are also able to impart their message about communal harmony.

The social development sector in India is largely governed by the objective of enabling marginalized individuals assert their due rights. But there are very few NGOs and individuals in this sector who have strategized their work with the purpose of addressing the issue of communal harmony. The following is indicative of the enormity of the problem. "According to the Home Ministry data, between 2001 and 2009, 6541 communal clashes occurred and 2234 persons were killed" (Engineer, 2011). Why does such staggering statistics not make it to the news?

The society is either too scared to bring it up due to the fear of getting the issue politicized which then leads to another wrath of communal clashes or the society has taken this form of violence in its stride and has become indifferent to it. If the former is true, it is indicative of the fact that there is still distrust and insecurity amongst the people and this implies that there is actually no peace between the religious communities. In order to find a pragmatic way forward we have to realise that the underlying issue is not just the riots *per se*, but also the process of how certain minority section of the society (Muslims) remain ghettoized and marginalized.

Why the government bodies are unable to reach out to them or why are those marginalized minority groups unable to come out and assert their dissent against the government's inability to cater to them? In order to look for answers one has to look deeper to see how the process of 'othering' gains momentum in day to day interactions and behind closed doors. Such alienating processes not only create further division, hate and distrust, but also impact the majority and minority groups in different ways.

Conclusion

By looking into various 'mediums of instilling communal harmony' being adopted in Mumbai and some other parts of the country, one gets to understand the spaces which are exploited by the political forces in order to instigate communal violence. There were two dominant forms of interventions for communal harmony. One was the 'workshop or lecture method' and the other a series of weekly group sessions with the same group of youth of a certain community with a history of communal violence. The former intervention is seen as a traditional form of engagement directed towards communal tension, as it aims to enhance and rectify knowledge regarding the group that is perceived as the 'other' and disconfirming stereotypes through information concerning broad range of members who are typically representative of the stereotyped group. However, academicians working on prejudice hold that these traditional forms of interventions are less effective when facing contemporary forms of prejudice practices. That is why, the individuals on the face value believe that they consider people from any ethnic background an equal citizen, but at the time of behaviour they resort to 'othering' practices. Fortunately, some interventionists are working with youth so as to equip them with skills that would make them more competent to operate in the society, and also create a sense of bonding within the group so as to enable them to engage in self-disclosure and appreciation of diverse views.

However, despite such novel and nuanced forms of communal harmony building interventions, the final vision of a nationwide social change is still difficult to achieve due to the adamant tendency of most of the NGOs working in the Indian social development sector who do not perceive communal harmony as an important social issue, even though they may be invariably witnessing 'othering' processes in the society frequently. The support of the government is unlikely except at the time of preventing riots because the government as well as other power structures, as some say, would less prefer to have a mass that is indeed critical and reflective.

The problem with letting this issue of communalism remain untouched is that such repressed and pent up hatred can spillover anytime, that in turn will again strengthen the process of 'othering' to make this whole process become even more vicious. And it is this acceptance of and indifference to overt and covert forms of communal violence that is simply unacceptable in a multi-cultural country.

REFERENCES

Dovidio, John F. Gaertner, and L. Samuel. 1999. 'Reducing Prejudice: Combating Intergroup Biases', *Current Directions in Psychological Science*, 8(4): 101-105.

Engineer, Asghar Ali. 2011. "Communal Riots in 2010," *Indian American Muslim Council*. http://iamc.com/op-ed/communal-riots-in-2010-by-by-asghar-ali-engineer-jan-3-2011-twocircles-net/ (accessed on 4th January, 2010).

Government of India (Sachar Committee Report). 2006. *Social, Economic and Educational Status of the Muslim Community of India,* Cabinet Secretariat, New Delhi.

Hamilton, D.L., and T.K Trolier. 1986. "Stereotypes and stereotyping: An overview of the cognitive approach", in J. Dovidio and S. Gaertner (eds), *Prejudice, discrimination, and racism,* pp. 127-163. San Diego, CA: Academic Press.

Joseph, Helen. 2000. "Salokha: A Response to Communal Riots, a Social Disaster", *Indian Journal of Social Work.* 61(4): 664-674.

Madan, T. N. 1992. "Secularism in its Place", in T. N. Madan (ed.), *Religion in India,* pp. 394-409. New Delhi: Oxford University Press.

Madan, T. N. 1992. *Religion in India,* New Delhi: Oxford University Press.

Mander, Harsh. 2009. *Fear and Forgiveness: The Aftermath of Massacre.* New Delhi: Penguin Books.

McEvoy, Siobhán. 2000. "Communities and Peace: Catholic Youth in Northern Ireland", *Journal of Peace Research,* 37 (1): 87-103.

Saussy, Carroll. 1993. "Pastoral Care and Counselling and Issues of Self-Esteem", in Robert J. Wicks and Richard D. Parsons (eds.): *Clinical Handbook of Pastoral Counselling,* pp. 363-365. Mahwah, New Jersey: Paulist Press.

Sherman, Jeffrey and Thomas Allen. 2011. *People With Low Self-Esteem Show More Signs of Prejudice.* http://www.psychologicalscience .org/index.php/news/releases/people-with-low-self-esteem-show-more-signs-of-prejudice.html (accessed on 3rd January, 2012).

Shourie, Arun. 1987. *Religion in Politics.* New Delhi: Roli Books.

Stearns, Peter N. 2011. "Why Study History", *American Historical Association: The Professional Association for all Historians.* http://www.historians.org/pubs/free/WhyStudyHistory.htm (accessed on July 11, 2008).

Weiner, Irving B. 2003. *Handbook of Psychology: Personality and Social Psychology,* Vol. 5. Sydney: John Wiley & Sons, Inc.

Williams, Angie, and Howard Giles. 1992. "Prejudice-Reduction Simulations: Social Cognition, Intergroup Theory, and Ethics", *Simulation and Gaming,* 23(4): 472-484.

Index

Academic examination of religious issues, 6
Affirmative action, 77, 85
Aga Khan Foundation, 99
Aga Khani Khojas, 64
Agricultural activities, 195
Agricultural University, 88
Agricultural workers, 5
Ahmadiyya movement, 4
Ahmadiyya Muslims, 4
Ahmedabad Municipal Corporation (AMC), 157
Ansari Muslims, 210
Anti-nationalism, 6
Anti-social elements, 77
Anti-sprawl development, 21
Art and Artifacts of Lucknow Chikankari, 178
Artifacts, 178
Artisan activities, 123
Arzal sections, 3
Ayodhya issue verdict, 227

Backwardness, 79, 179, 201
Belief systems, 44, 50
Bharatiya Janta Party (BJP), 219
Bichoulias act, 180
Bihari Muslims, 64, 120
Blighted areas, 23
Bohra Muslim, 112
Bombay Port Trust, 96
Bombay Presidency and Province, 62
Brihan-Mumbai Municipal Corporation, 71, 120
British policy, 4
British rule, 4
Broad Usual Activity Status in Mumbai, 137
Broken window theory, 32–33
Burqa practice of, 54–55

Capitalism, 20, 23, 39, 177, 182
 global, 21
 urban, 39
Capitalist system, 20
Caste differences, 83
Catholic Educational, 88
Centralized manufacturing processes, 181
Cheeta Camp, 145
Chikankari workers, 14, 177, 187, 189–90, 192, 196
Child healthcare practices, 195
Christian terrorism, 220
City Development Plan, 35
Civic amenities, fundamental, 212
Civil society, 86, 112, 116, 172, 177
Cloth manufacturer/trader, 216
Committee for Protection of Democratic Rights (CPDR), 132

Communal conflicts, 15, 233
Communal harmony programmes, 236
  issue of, 240–241
Communal riots, 116, 157, 211, 221, 226, 233
Communalism, 209, 226, 235, 237
  issue of, 238, 242
Concerted public action, 156
Cooperative interdependence, 46
Crime mafia, 22
Cross-cultural conflicts, 44
Cultural
  activities, 49
  associations, 200
  matrix, 175
  systems, 175

Dakhni Muslim, 210, 219
Demographic Characteristics, 158, 168
Deprivation of Muslims, 14, 78
Dirty water, 73
Discriminatory practices, 123
Disruptive activities, 26
Distance education, 87, 89–91
Drinking water, 148, 159, 162, 215, 217

Economic activities, 166, 218, 222
Education
  female, 83, 103
  formal, 142, 198
  parental, 14
  primary, 142, 172
  professional, 98
  secondary, 87
  technical, 14
  value, 51
Educational attainments, 10, 78, 83–84, 94, 142
Educational Deprivation, 169
Educational Equality, 82
Educational Status of Muslims in Mumbai, 65
Elite groups, 155
Employability, 141
Employment
  formal, 138, 218
  low-paid, 6
  public, 188
  regular, 188
Employment/Unemployment, Categories of, 136
English medium education, 95
Ethnic
  background, 242
  groups, 21
  identity, 43–44
  minorities, 153
  segregation, 19, 35–36, 39
Ethnocentric policy, 8
Ethno-cultural groups, 45
Extra-curricular activities, 113

Factory system, 181–82
Familial skills, 175
Female education, 104
Female workers, 140
Feudal society, 155
Foreign Muslims, 3
Form labour unions/associations, 184
Formal institutions, 130, 149
Friday prayers, 53, 211
Fundamental components, 228
Fundamental issues, 209

Gandhi, Indira, 5
Ganesha festival, 74
Garbage, 74–75
Garbage disposal mechanism, 74
Gentrification, 23
Ghettoisation, 116–18, 145
Giri Institute of Development Studies, 196
Gopal Singh Committee Report, 5
Government welfare programmes, 2

Group-occupation nexus, 200

Haitian women, 57
Handloom-weaving centre, 214
Hindu fundamentalist movements, 41
Hindu nationalism, 42
Hindu rightist groups, 157
Hindu-dominated villages, 220
Hinduism, 235–36
Hindu-Muslim
  Riots and Participants, 229
  unity, 4
  violence, 41
Horizontal schisms, 18
Household
  industry, 193, 199
  poverty, 125, 127
Human right activists, 23
Hunter Commission Report, 5
Hyper-punitive practices, 8

Ideological mechanisms, 15, 176
Immigrant groups, 44, 58
Impoverished Muslims, 115
Indian Muslim Congress Party (IMCP), 220
Industrial areas, 214
Industrial economy, 115
Industrial revolution, 116, 181
Industrial riots, 116
Industry
  chikankari, 175–80, 182–83, 188, 195, 198, 203
  cloth/textile, 214
  garment, 199
  hammock, 176
  power-loom, 15, 209, 212, 215–16, 221–22
  textile, 211
Informal sector activities, 6
Informal water market, 73
Institute of Distance Education (IDE), 86–90
Institutional credit, 130, 142, 145, 149
Institutional legitimacy, 46
Institutional policy, 173
Intelligence department, 212
Interactions, 8, 56, 58, 88, 117, 187, 236, 238
  cross-cultural, 109
  diverse inter-group, 240
  human, 227
  inter-community peer, 110
  inter-religious, 238, 240
  peer, 110
  personal, 115
  positive, 58
Inter-community harmony, 226
Intergroup conflicts, 46
International Labour Organisation (ILO), 153
Islamic practice, 54
Islamic terrorism, 6

Jamia Millia Islamia University, 8, 47, 49
Jamia Nagar, 8, 47
Jan Swaraj Party (JSP), 220
Janata Party, 219
Jharkhand, 24
Joint families, 195
Judicial system, 35
Justice Rajender Sachar Committee Report, 28

Kolkata Municipal Corporation (KMC), 157–58
Konkan Municipal Councils, 125
Konkani Muslims, 64, 115

Labour union retaliation, 199
Liberal practices, 55
Loom-workers, 215–16, 219
Lucknow and Network of Entrepreneurship and Economic Development, 194

Lumpen elements, 116

Mahajans, 179, 193–94, 199–200, 202–3, 205
Maharashtra Navnirman Sena (MNS), 220
Majority-minority conflicts, 20
Malabar Hills, 33
Malegaon Municipal Corporation, 210
Malegaon, 209–22
  civic amenities, 211
  communal riots, bomb-blasts, 211–12
  education and employment, 216–18
  health infrastructure, 213–14
  housing and sanitation, 213
  politics, 218–220
  powerloom industry, 214–16
  tryst with history, 209–10
Maliwadi, 210
Marginalisation of Muslims, 5, 14
Marginalised
  families, 149
  sections, 26
  groups, 195
Marriage and Social Interaction, 107
Media sector, 82
Mid-day meal, 217
Migrant Muslims, 114
Mill workers, 20
Minority section, 241
Mohalla factor, 106
Motivational factor, 119
Mufti Mohammed Ismail, 220
Multicultural models, 45
Multiculturalism, 45
Mumbai Metropolitan Region Development Authority (MMRDA), 127, 146
Mumbai Urban Transport Project-II (MUTPII), 127
Municipal Corporation of Greater Mumbai (MCGM), 35, 85
Muslim Chikankari, 14
Muslim concentration areas/ghettoes, 145
Muslim education trust, 88
Muslim ghetto, 64
Muslim ghettoes, 36
Muslim jail inmates, 26, 28
Muslim migrants, 47
Muslim nationalism, 42
Muslim Personal Law, 108
Muslim prisoners, 28
Muslim rule, 41
Muslim traditions, 104
Muslim Trust College, 108–9
Muslim weavers, 115

National Crime Report Bureau, 24
National Sample Survey Organisation (NSSO), 1, 125, 135–36
Navi Mumbai Special Economic Zones, 23
Neoliberalism exacerbates, 19
Neoliberalism, 8, 18, 21, 24, 32, 38–39
  opponents of, 21
  spatiality of, 21–22
Nizam-ruled state of Hyderabad, 63
Non-economic factors/institutions, 177
Non-Governmental Organizations (NGOs), 114, 121, 156, 194, 229, 234, 236–40, 242
Non-institutional sources, 130
Non-marginalized sections, 121
Non-Muslim areas, 212
Non-Muslim Educational Trust College, 101

Okhla, 47
Organisational pyramid, 179–80

Organised Crime Act, 26
Organizational pyramid, 179
Organized factory, 193
Organized forums, 99
Other Backward Classes (OBC), 9, 124–25, 127–29, 131–35

Park Circus, 157–58
Peer-Group Interaction, 110
Peripheral manufacturing units, 23
Phobias and Fears, 37
Physical education teachers, 238
Pluralistic policy, 85
Political awareness, 49
Poverty line, 10, 63, 65, 125, 127, 142, 149, 196
Poverty
  city level family, 127
  family/household, 127
  incidence of, 10
  measuring, 127
  rampant, 137
  removing, 24
  rural, 63
Praja Socialist Party (PSP), 219
Prevention of Terrorism Act (POTA), 26
Public sector companies, 92
Punishment practices, 19
Purchasing power, distribution of, 154
Purdah system, 15, 51–52, 54–56, 113, 175–76, 180, 183–84, 186–88, 194, 197–99
Purdah-Poverty Nexus, 184

Quota, 95
Quran, 51, 55–56, 94–95, 99, 109–10

Rabble-rousing, 219
Racial/ethnic cleavages, 153
Ranganath Misra Commission Report, 5
Reactionary practice, 56
Red zone, 9
Reductionism, 177
Religious
  conflicts, 8
  education, 50–51, 53
  festivals, 50
  institutions, 44
  organizations, 44
  practices, 49, 52–53, 227
  war, 220
Research and Action Group, 132
Right wing nationalism, 235
Rioters, 34, 232
Ritual practice, 53

Sachar Committee Report, 6, 9–10, 41, 63, 82, 188, 227
Salaried employment, 103
Sameera Khan, 64
Saqib Khan, 15
Scattered domestic industries, 182
Second World War, 32
Sectarian
  differences, 115
  groups, 37
  politics, 37, 123
Sectors
  handicraft, 176
  informal, 166, 175, 187, 218
  organized, 75
  private, 75, 77–78, 149
  public, 78, 150
  textiles, 182
  unorganised, 75, 78
  welfare, 233
Secularism, 227–28
Self-actualization, 44
Self-employed families, 142
Self-Employed Women's Association, 194
Self-empowerment, 86
Sellers/bigger contractors, 180
Sewage system, 73–75

Shah Bano's case, 108–9
Shatabdi/Centenary Hospital, 71
Shia sect, 124–25
Shiv Sena, 220–221
Shrikrishna Commission Report, 34
Sidha ilzaam Muslims, 232
Sindhi and Gujarati Society Educational, 88
Slum dwellers, 22–23, 152, 162, 164, 172
Slums/chawls, 23, 65, 67, 146, 159
Social awareness forum, 99
Social change, 204–5, 234–35, 242
Social conflicts, 8, 38
Social empowerment, 155
Social exclusion, 153–56, 159
Social interaction, 84–85, 107
Social practice, 187
Social Structures of Accumulation (SSA), 177
Societal institutions, 45
Sociological characteristic, 156
Socio-political factors, 84
Socio-Religious Categories (SRCs), 1, 6, 9–10, 83
Socio-religious groups, 1, 125
Spatial Exclusion of Muslims in Cities, 156
Special Economic Zones (SEZs), 23
State-society relationship, 39
Stereotyped groups, 242
Sub-contracted washermen, 180
Sub-contracting system, 15, 182, 203
Suburban development, 32
Sunni Muslims, 41, 112

Talaq, 108, 110
Tata Institute of Social Sciences, 127
Teesra Mahaz (Third Front), 220
Terrorism, 6, 21, 37
Terrorist activity, 232
Thane Municipal Corporation (TMC), 65, 68, 72, 74

Ultra-sensitive zone, 212
Undertrials, 26, 28
University of Mumbai, 86–89, 100, 119
Untouchable caste, 3
Urban poverty, 63, 159
Urbanism, 32–33
Urdu medium schools, Availability of, 95
Uttar Pradesh Export Corporation Limited (UPECL), 204

Viceroy Lord Mayo, 5
Violence, mob, 220
Vishwa Hindu Parishad's Dharmayudh (religious war), 220

Wage employment, 98
Wahabi movement, 4
Water Connection and Electricity, 197
Water infection, 196
Water pollution, 74
Water shortage, 148
Water-mafia, 73
Well-to-do families, 103
Women
- artisans, 204
- burqa clad, 118
- divorced, 106
- embroiderers, 184
- representation, 84
- rights literacy, 121
- role of, 52, 59
- work participation of, 102, 149

Women's Research and Action Group (WRAG), 132
Workers
- child, 137
- earnings of, 141
- home-based, 76
- jari, 92, 98
- lace, 199

loom, 216, 222
migrant, 210
part-time, 193
powerloom, 216, 219
salaried, 188
semi-skilled manual, 136, 140–141
skilled manual, 141
unpaid family, 6
unskilled, 121, 123, 198
Workshop or lecture method, 241
World Bank, 22, 127

Xenophobic society, 188

Yarn manufacturers, 216, 222

Zero tolerance, policy of police, 32